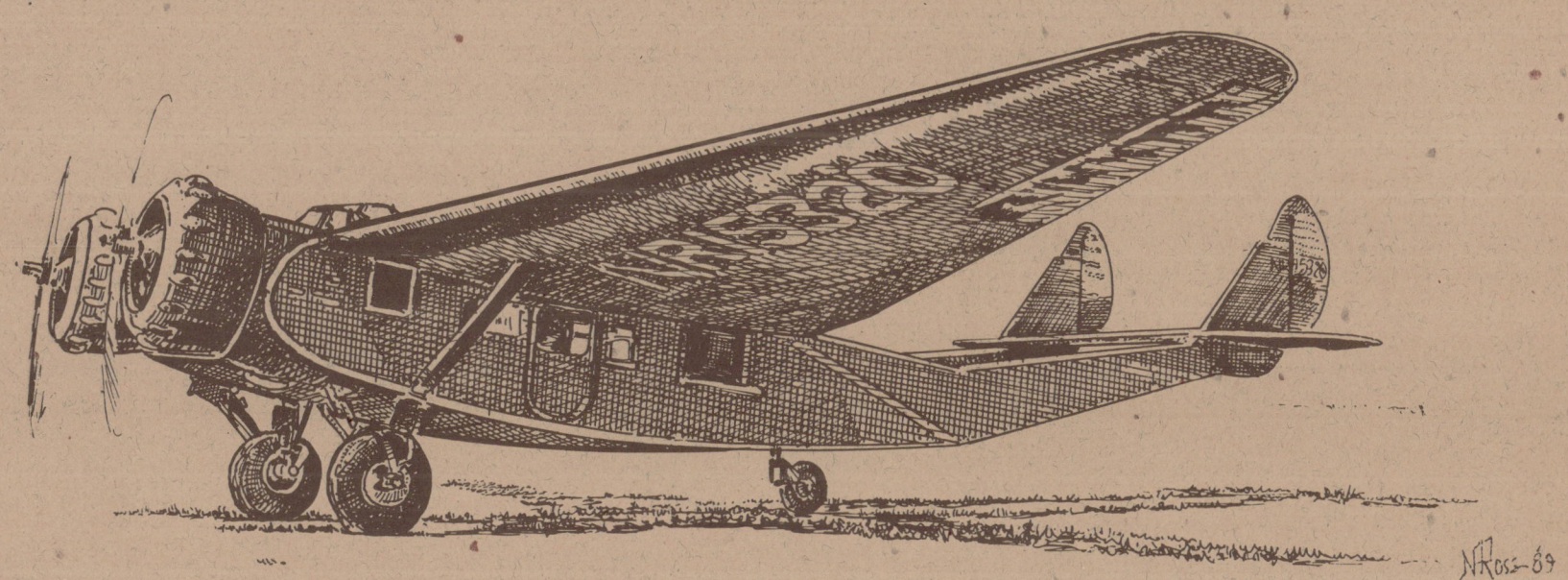

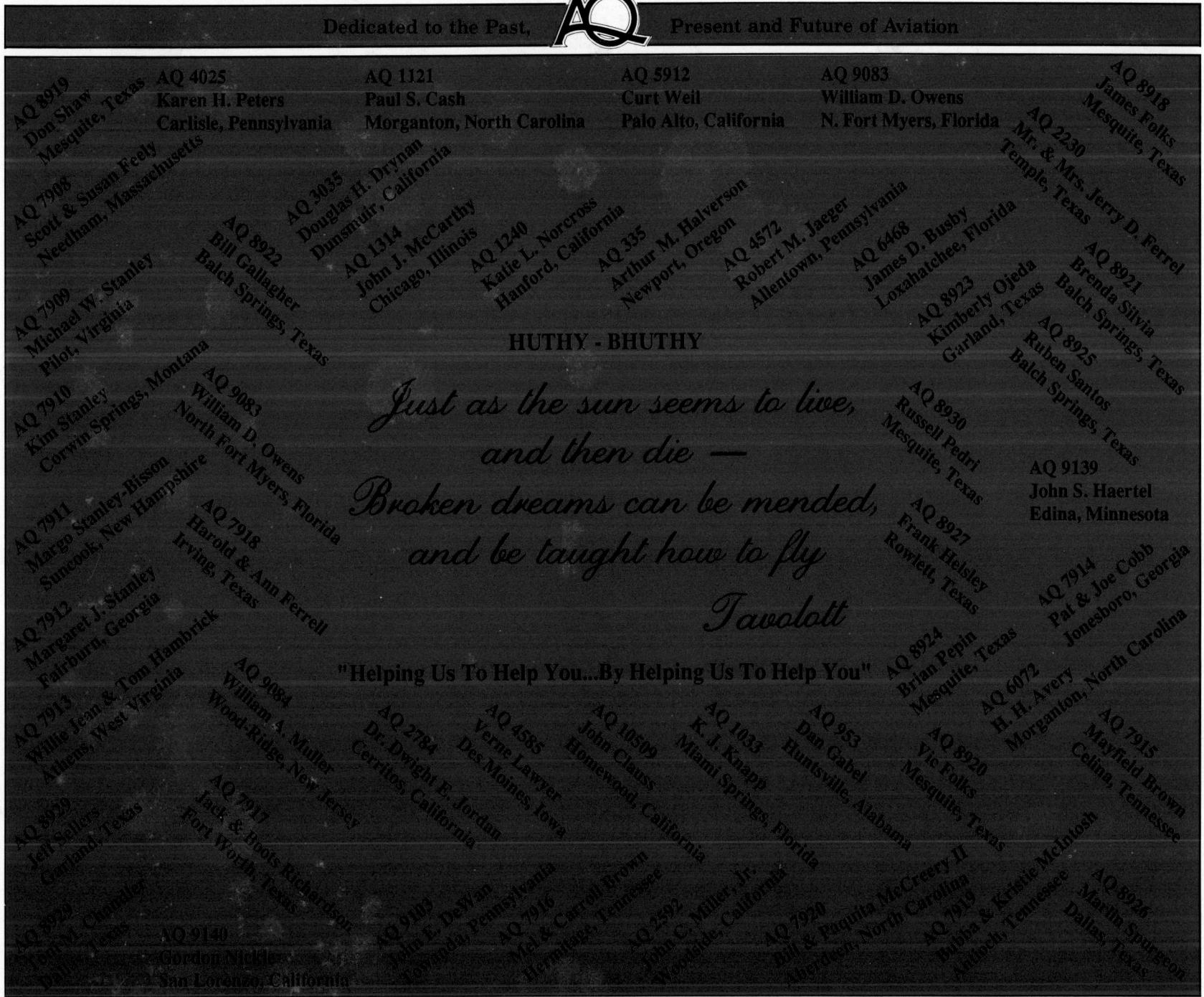

AQ
AVIATION QUARTERLY

The Enthusiasts' Magazine of Aviation

Volume Nine   Number Two

Publisher
George E. Stanley

Editor
Frank B. Thornburg, Jr.

Customer Service
Anna C. Dawn

Artist
Norman A. Ross

Graphic Arts
Marlin Spurgeon

Typesetting
Toni M. Chandler

Frontispiece	**Front End Sheets**	**Back End Sheets**
Fledgling	Left: Curtiss NC 868N	Left: Fairchild "Trainer"
Tail Assembly.	Right: UB-14B	Right: Convair YB-60

AVIATION QUARTERLY

SPRING 1989
CONTENTS

129 • HUTHY BHUTHY: HELPING US TO HELP YOU.
By helping us to help you. *by AQ Charter Members*

132 • DOUGLAS H. DAVIS
A Georgia Cracker with few peers *by Clair C. Stebbins*

150 • 1946 AERONCA CHAMP TAIL DRAGGER
Too Few remain. *by Raymond Murray*

152 • VINCENT J. BURNELLI
Results were many, accolades too few. *by Kathryn Jones*

168 • CURTISS FLEDGLING NC 868N: AN AQ COLOR PICTORIAL
In Oregon, one of four in the world. *by Dick Hopkins*

178 • WILLIAM E. KEPNER: ALL THE WAY TO BERLIN
Makers of the United States Air Force. *by Paul F. Henry*

208 • DICK MERRILL, FREDERICKSBURG AND THE V. A. M.
A compendium relating to aviation in Virginia. *by Yvonne C. Pateman*

223 • THE NEW-DAY BARLING NB-3
Setting new standards in production. *by Company Brochure*

228 • ELWOOD R. QUESADA: TAC AIR COMES OF AGE
Makers of the United States Air Force. *by John Schlight*

253 • JUST PICTURES
What could have been to what was. *by AQ Editors*

256 • AQ RECOMMENDS: THE WAR IN SOUTH VIETNAM
The Years of the Offensive 1965 - 1968 *by John Schlight*

AVIATION QUARTERLY

In helmet, goggles and leather jacket, Doug Davis in typical attire of mid-1920s barnstormer.

DOUG DAVIS

Air Racer, Barnstormer, Airline Pilot

by
Clair C. Stebbins

Coney Island's sandy beach was jammed on the sultry Fourth of July in 1926. Newspapers estimated the number at 600,000, a crowd that included Governor Al Smith, who strolled the boardwalk shaking hands with his holidaying constituents while photographers took pictures that would appear in the next Sunday's rotogravure sections.

Suddenly overhead there appeared a low-flying biplane from which tumbled hundreds of tiny parachutes, each bearing a small candy bar in a red and white paper wrapper. This set off a wild stampede as the bathers scrambled for a share of the sweets. It is not recorded that the governor or any of his party took part in the stampede as the airplane circled and unloaded more parachutes along the entire length of the beach. In a first-page report on the city's holiday observance the next morning, *The New York Times* described the incident this way: "The crowd on the beach at Coney Island was nearly stampeded by an airplane which flew over the water front, dropping sample boxes of candy. There was a wild scramble for the sweets by men, boys and women. Mrs. Yetta Kerman, 54 years old, of 164 South Fourth Street, was knocked down and her left leg was broken in the mad rush for the candy. She was taken to the Coney Island hospital."

Soon after his historic flight to Paris, Col. Charles Lindbergh (left) visited Davis (center) in Atlanta. Pictured with them is one of Davis' pilots.

The pilot was not identified but his name would appear in *The Times* and the newspapers all over the country many times in the next eight years. He was Doug Davis, a barnstormer from Atlanta whose name was written large in the skies over America during aviations glamour years, spanning the decade and one-half following World War I. Among the popular heroes in leather jackets, whipcord breeches and white scarfs, he shared headlines during this period with such other flying immortals as Jimmy Doolittle, Roscoe Turner, Wiley Post and Charles Lindbergh, and their female counterparts including Amelia Earhart and Jacqueline Cochrane.

Douglas H. Davis, who grew up on a farm in central Georgia, was a high school senior in 1917, the year the United States entered World War I. Without waiting for his diploma, he enlisted in the Army Air Service with visions of participating in the aerial dog-fights over France that were described so vividly in newspapers and magazines of that day. This was not to be, however. He graduated from army flight school at the head of his class but when he asked how soon he would be going overseas, a superior officer told him he wasn't going. "There's a shortage of capable instructors and you're needed here to teach others to fly." A disappointed Lieutenant Davis spent the next year in the instructor's cockpit of a Curtiss JN-4D trainer, the famouse spruce and fabric biplane better known as the Jenny. Like many other military pilots when the war was over, he seized the opportunity to buy a surplus Jenny from the government for a few hundred dollars.

From a pasture field near his hometown of Griffin, Ga., with occasional visits to nearby towns, he made a tidy living the next summer by taking passengers for ten-

minute rides at five dollars each. Sensing potential profits in the new field of aviation, he moved to Atlanta and started his own flying service, building a hangar at the Candler race track and persuading an Ohio airplane manufacturer to grant him the Georgia distributorship for Waco airplanes.

He sold his Jenny and bought three Wacos which he used as demonstrators during the week and for barnstorming excursions throughout the southern states on weekends, hiring former army pilots to join in what now became the Doug Davis Flying Circus. In addition to his own demonstration of aerial acrobatics, his thrill show included a wing-walking exhibition by a glamorous daredevil, Mabel Cody, billed as the niece of "Buffalo Bill" Cody.

About this time, Doug Davis' blossoming career as an aviator crossed paths with the career of a Chicago candy-maker, Otto Y. Schnering. It was a fortunate coincidence for both. Schnering not only knew how to make tasty confections; he knew how to sell them, and was recognized as one of America's most imaginative and promotion-minded business men. As president of the Curtiss Candy Company, he had developed a chocolate-coated candy bar that combined roasted peanuts, corn syrup, coconut and a half-dozen other ingredients. He needed a catchy name and a novel means of introducing it to the public.

Babe Ruth was the best known name in the country — better known even than President Harding or General Pershing — but Schnering had no intention of paying royalties to the baseball idol for the use of his name. So he did the next best thing, according to oft-told legend. There happened to be a baby named Ruth among Schnering's relatives so he professed to be naming the candy bar in her honor. Thus the five-cent Baby Ruth candy bar, with a red and white wrapper, was born in the Curtiss candy kitchens in Chicago.

Next, he conceived the idea of harnessing the nation's love of baseball to the growing popularity of the airplane. What better way to introduce Baby Ruth candy than to have it dropped from the skies? His search for a reliable pilot for his promotion stunt led him to Atlanta and Doug Davis. They made a deal and the name, Baby Ruth, was painted in large bold letters on all three of the Davis Wacos and the barnstorming operation took on the name of the Baby Ruth Flying Circus.

Wherever the circus appeared, the candy bars were dropped on the assembled crowds, each bar attached to a small parachute made of rice paper. What happened in the next three years, when the showmanship of Otto Schnering combined with

On their wedding day, Doug Davis and his bride, Glenna Mae, set off for their honeymoon in one of his "Baby Ruth" biplanes.

the airmanship of Doug Davis, was an advertising blitz unlike anything the country had ever seen. The Baby Ruth airplanes ranged over 40 states, dropping candy on county and state fairs, racetracks and bathing beaches — wherever crowds were gathered.

A Curtiss publication described what happened in Pittsburgh one day in 1923:

> "It was an ordinary day. Automobiles were criss-crossing the business district noisily, people were walking on the sidewalks and others were at work in plants and offices. Over the city, an aviator named Doug Davis glanced down with a sly smile.
>
> "Within five minutes, Pittsburgh was a screaming pandemonium. Doug Davis was roaring through the business district, a few dozen feet above the streets, looping, rolling, and swooping between buildings. Upper-floor office workers experienced the sensation of hearing a roar, glancing out the window and seeing a flying machine a few feet away. People rushed to the streets and rooftops. Children ran out of schools, drivers left cars standing at intersections. Traffic was jammed for blocks and blocks and blocks.
>
> "When he was absolutely certain that he had Pittsburgh's attention, Mr. Davis did what he'd been hired to do. He climbed to the proper altitude and dumped his cargo overboard. Down upon Pittsburgh rained hundreds of tiny parachutes — each carrying a five-cent Baby Ruth bar. The pandemonium was doubled. People risked falls from windows reaching for the 'chutes. Children ran out into the streets (without danger — traffic was hopelessly snarled) and adults fought for the free candy.
>
> "After a couple of hours, when Davis had run low on gas and had run out of Baby Ruths, Pittsburgh was restored to some sort of order. It was at once obvious to the city fathers that, unless something was done quickly, Pittsburgh would be in constant turmoil, free candy or no free candy. The city council met in emergency session and passed an ordinance prohibiting airplane flights over the city below a height of several hundred feet and making candy bars on parachutes illegal."

Making allowance for exaggeration in the writer's account of the airplane's low-flying caper, this and other stunts created newspaper headlines from coast to coast, which was something the crafty Mr. Schnering had anticipated.

In Milwaukee, a woman broke a leg in the mad scramble for Baby Ruth bars. She sued Curtiss and collected. A traffic policeman in Cincinnati was fired for leaving his street corner post to chase parachutes. In his defense, he said traffic was so hopelessly snarled that his services were of no use. When Davis flew into Miami to spill candy bars on Hialea racetrack, he found a ten-year-old volunteer to pitch the candy bars from the cockpit. His name was Paul Tibbets, whose father operated a wholesale confectionery business and was southern Florida's principal distributor of Baby Ruth candy. It was the first airplane ride for the boy who, 20 years later, would drop the world's first atomic bomb in Hiroshima. In his autobiography, *The Tibbets Story*, retired Brigadier General Paul W. Tibbets observes that his peacetime "bombing" of the racetrack was a more memorable experience than his historic mission to the Japanese city.

In every city and town he visited as a barnstormer, handsome Doug Davis was flirtatiously pursued by the flappers for whom the post-war decade is remembered. However, friends recall that his sole romantic interest was a girl back home in Atlanta — Glenna Mae D'Hollosay. They corresponded while he was in service and resumed dating after the war. His first plane was named for her, with *Glenna Mae* painted in bold block letters below the pilot's cockpit. They were married on Christmas Day, 1925, and promptly set out on a honeymoon in one of the Baby Ruth airplanes.

Meanwhile, behind a small desk in one corner of his cluttered hangar at Candler, the pilot was becoming a hardheaded businessman. He was making money with the sale and repair of planes. The weekend barnstorming forays were also earning a profit, sometimes grossing as much as $3000 in a single day. He became known throughout the South as the "king of the barnstormers," his only serious rival being Ivan Gates, a colorful cigar-chewing operator of a flying circus of similar size.

More powerful airplanes were being developed and Davis dropped his Waco dealership and became the regional distributor for the Travel Air, built at Wichita, Kansas. With one of these planes, he began entering air races, a sport of growing popularity. His success was such that Travel Air engaged him to fly a newly developed plane in the free-for-all highlight event of the National Air Races at Cleveland in 1929. Such secrecy had surrounded the building of the plane that newspaper and magazine writers, being supplied with tantalizing fragments of information about its design, began calling it "the Travelair Mystery Ship."

On the first day of the races, Davis flew

The red and black "Mystery Ship" in which Doug Davis won the first Thompson Trophy race at Cleveland in 1929.

AVIATION QUARTERLY

Doug Davis with trophy he won in the first Thompson Trophy race at Cleveland in 1929.

the R613K with the Chevrolair engine to a first-place finish in an event for experimental aircraft. Later the same day, he led a four-man team to victory in a relay race, flying a Travel Air biplane powered by an OX-5 engine. He and his team had to settle for second place, however, in acrobatic competition.

By the time of the Labor Day climax, interest among aviation people was at a peak over the still-hidden Mystery Ship that Davis would fly in the free-for-all for planes of unlimited horsepower, an event for which Charles E. Thompson of Thompson Products had offered a large silver cup and $1500 in prize money. Even so, the military service were the favorites, having dominated the free-for-all event in previous years. A duel for first place was predicted between Army Lt. R.G. Breene in a Curtiss P3A pursuit plane and Navy Lt. Cmdr. J.J. Clark, flying a Curtiss Hawk. Davis was one of five civilian pilots entered in the 50-mile race over a ten-mile triangular course.

Lt. Breene was off to a fast start but, by the time the first pylon was reached, Davis and the Mystery Ship had moved ahead. Then for the next two laps, the Georgia barnstormer stretched his lead until, at the start of the third lap, he cut inside one of the pylons, an error that would disqualify him unless he turned back at once and re-circled the red and white tower. Without hesitation, he put the little monoplane into a vertical bank and spun around the pylon, a maneuver that put him back in the race but no longer in the lead. With 30 miles to go, he pushed the little red and black monoplane and its Whirlwind engine to their limit, passing the other competitors including Lt. Breen and Roscoe Turner, to cross the finish line two miles in front of the army plane and eight miles ahead of Turner. His speed of 194.9 miles an hour set a new closed-course record.

The 1929 National Air Races are remembered today as a milestone in aviation history because they marked the first time that a civilian airplane had bested the military.

When Davis climbed out of the cockpit to accept the Thompson Cup from actress Mary Pickford, he was wearing a white shirt and tie. Except for the essential helmet and goggles, his attire was a departure from what the public expected of its air heroes who, for the most part, wore flashy leather jackets and silk scarfs. In fact, Roscoe Turner never gave up the uniform and for him this included whipcord breeches, polished leather boots and a military-style jacket with the airman's wings embroidered above the left breast pocket. Turner also had a lion cub named Gilmore for a companion on many of his cross-country flights, but that is another story.

Less than two months after the 1929 races, Travel Air merged with Curtiss-Wright and Walter Beech moved to New York as the company's vice president in charge of sales. Davis became manager of Curtiss-Wright's southern territory. With this new commitment, he and his Atlanta business partners sold a small air service they had started under the name of Davis Air Lines to Texas Air Transport. Under Davis' management, it had operated between Atlanta and Birmingham. In the

celebrated Mystery Ship, now wearing the Curtiss-Wright label and no longer a mystery, Davis made exhibition flights around the country and broke many speed records, among them a four and one-half hour flight from New York to Atlanta.

When Harold Pitcairn began flying the mail between New York and Atlanta in 1928, he recruited pilots in Georgia, most of them former barnstormers. Davis was approached with what he conceded was a tempting offer, but he chose to remain with his successful flying service. However, when Pitcairn's operation was transformed into a passenger-carrying service under the name of Eastern Air Transport, he found it impossible to turn down an offer to become Eastern's premier captain. The airline reaped dividends in the form of newspaper headlines, for the name of Doug Davis was already well known in the word of aviation.

On December 10, 1930, Davis was at the controls of a new Curtiss Condor on its initial scheduled flight from New York to Atlanta, with stops at Philadelphia, Baltimore, Washington, Richmond, Greensboro, Charlotte, Spartanburg and Greenville. A large crowd was on hand at Candler Field, now the city-owned airport, when Davis landed after a trip of seven hours and 45 minutes, a time that included the eight stops with a 30-minute pause for lunch at Greensboro. The Curtiss Condor was an 18-passenger biplane with sound-proofed cabin that, it was claimed, reduced the interior noise level to that of a Pullman railroad car. Powered by two 12-cylinder engines, its top speed was 147 miles an hour.

In uniform of Eastern Airlines captain, Doug Davis is pictured here with Curtiss Kingbird airliner.

AVIATION QUARTERLY

A victory wave from Doug Davis and congratulations from Vincent Bendix at end of 1934 Bendix Trophy race from West Coast to Cleveland.

Because of his fame as a pilot, Davis was assigned to other inaugural flights as the airline expanded its operations. The story is told of an 80-year-old woman passenger who flew from Atlanta to Miami on one of his early trips. It was her first airplane flight and, as was his custom, the pilot took time to greet every passenger personally. When it came time for her to return home, the woman politely informed the airline reservations clerk that she would travel only if Davis was flying the plane. She was booked on his next flight, even though it meant for her a two-day delay.

While Eastern did not hesitate to capitalize on its most famous pilot's reputation as an air racer and acrobatic headliner, the company may have had occasional misgivings. At a time when airlines were striving to convince a skeptical public that flying was a safe and dependable way to travel, they had a nationally-famous daredevil in the cockpit. Nevertheless, Captain Doug Davis the airline pilot bore little resemblance to Doug Davis the air racer and stunt flier. The death-defying instincts that drove him to new speed records and hazardous aerial acrobatics disappeared when he took the controls of a Condor or Kingbird. He flew airliners strictly by the book.

Weekends frequently found him taking off in one of his own planes for some distant city where an air race was being held. When there were no races, he would practice acrobatics by the hour in his special-build NC612-K Wright-powered Speedwing Travel Air.

At Chicago's National Air Races in 1930, he did not enter the Thompson free-for-all,

Doug Davis with numerous trophies won in 1930s acrobatic and racing entries.

The Gee Bee R-1 racer of 1932, built by the Granville brothers in Springfield, Massachusetts, established a new world's speed record for landplanes at 296.28 mph. Jimmy Doolittle was the pilot.
(Thompson Products Company)

Matty Laird's "Super Solution" of 1931 averaged 223 mph between Burbank, California, and Cleveland, Ohio, in the trans-continental Bendix Trophy race with Jimmy Doolittle as pilot.
(E. M. Laird)

but took part in the daily aerobatic competition and, at the end of the week, was awarded first prize for his rolls, loops and spectacular tumbling performances. In a borrowed plane that week, he finished second in a balloon-bursting contest, an event for which he had never practiced.

Fate gave Doug Davis his next chance at air racing glory in 1934 when he was invited to fly the fastest of three Wedell-Williams planes in the Bendix cross-country event and the Thompson classic at the National Air Races in Cleveland. Jimmy Wedell, the plane's co-designer, had been killed in a crash a short time before. Davis eagerly accepted and obtained a leave of absence from his duties at Eastern Air Lines (the new name for Eastern Air Transport). He picked up the racing plane and flew to Los Angeles to compete in the Bendix dash to Cleveland.

"I'm going to win enough money to buy a farm," he told a friend, John Howard, before his departure from Atlanta.

The Wedell-Williams "44" was an air race pilot's dream plane. On the fuselage was painted the boast, "Hot as a .44 and Twice as Fast." Despite its performance capabilities, the plane was known to develop a wing flutter when the engine was at full throttle. On the Bendix flight from Burbank, Calif., to Cleveland, Davis encountered trouble with the stabilizer trim as he battled turbulence and rough weather. After stops at Goodland, Kansas, and Lansing, Illinois, he flashed across the finish line 36 minutes ahead of J.A. Worthen, flying another Wedell-Williams plane, to the cheers of 34,000 spectators at Cleveland airport. He was presented with a trophy by movie star Mary Pickford and a $5400 check by Vincent Bendix.

In the week of racing that followed, Davis won three more events and thrilled spectators with a 306.2 mile-an-hour straightaway speed dash. As the day of the Thompson Trophy race approached, a duel was anticipated between Davis and Roscoe Turner who, after losing the Bendix race, had continued to New York, setting a new coast-to-coast record of 10 hours, 2 minutes and 57 seconds.

At a breakfast for pilots on the morning of the Thompson, Davis was critical of the race management for having shortened the triangular course from ten to eight and one-third miles in order to give spectators a better view of the planes in their 12 low-level circuits of the three pylons, each requiring a steep turn of 120 degrees.

"Someone may get killed this afternoon," he said, observing that the course was too short for high performance planes flying at low altitude.

Eight planes took off at the start of the race, with Davis pulling in front as they left the scattering pylon, and Turner in close pursuit. The Wasp engine of No. 44 was performing perfectly and Davis had stretched his lead to a little more than a mile as he entered the eighth lap, almost two-thirds of the way through the race. At that moment, misfortune struck as he drifted inside the No. 2 pylon. At high speed and low altitude, the red and white checkered pylons become a blur to the racing pilot, looking through heavy goggles from an open cockpit with the wind beating against his face at every turn. The same mistake had cost Turner the race the previous year.

There was no choice but to re-circle the missed pylon, a time-consuming maneuver that would put his rival in the lead, but Davis was confident that he could catch up in the five laps remaining. After all, his plane had outsped Turner's by 11 miles an hour in the straightaway speed test two days before. To make the turn with a minimum of lost time, Davis whipped the plane into an almost vertical bank from which he never recovered as the fast little craft was thrown into a high speed stall and spiraled to the ground. The heroic pilot perished instantly in a mass of flames but the crash occurred out of view of the 60,000 spectators who were told by the race announcer that he had been forced to bail out.

Turner went on to win the race and the

Walter Beech flew this Travel Air to first place in the 1928 Ford Reliability Tour. *(Beech Aircraft Corporation)*

$4500 first prize at a speed of 248.129 miles an hour but there was no victory smile on his face as he accepted the handsome silver trophy from Fred Crawford, president of Thompson Products. Doug Davis was one of the best-loved figures in aviation and the flying fraternity had lost a hero.

There is a postscript to the Doug Davis story, also tragic. Doug Davis Jr., who was six years old at the time of his father's death, served two years with the army in Europe after World War II. His interests turned to art rather than aviation. In 1949, he graduated from the Atlanta Art Institute, after which he studied at the Sorbonne in Paris under the GI bill.

He specialized in portrait and figure painting and was commissioned to paint portraits of many socially prominent people on both sides of the Atlantic, including Mrs. Walter Chrysler, the Countess Grace d'Aatier, and the Belgian princess Elizabeth de Croy. He also painted a number of murals, both overseas and in this country. One is still to be seen in the American service club at Frankfurt, Germany, and another at the Surf Club in Miami.

In the mid-fifties, he opened a studio in Paris and his work soon came to be admired by the international community on the continent, where he was regarded as one of the world's most promising young artists. He traveled in artistic and literary circles with a coterie of friends that included Edith Piaf, the Parisian Chanteuse, and the American poet-philosopher, Rod McKuen.

He was approaching the peak of a successful art career when, in 1962, a group of 105 Atlantans, including many civic leaders, took a charter flight to Europe to visit museums and points of cultural interest. Among those making the trip were a number of Doug Junior's hometown friends who persuaded him to return with them for a visit, inasmuch as there was an empty seat on the plane. All lost their lives when the Air France Boeing 707 crashed on takeoff from Paris' Orly Field on June 3, 1962.

Doug Jr. was 34, one year younger than his father when he was killed. His friend McKuen remembered him with a number of poignant verses.

Today, Glenna Mae Davis lives with her memories of two airplane crashes that claimed her husband and son. On the walls of her home on Georgia's St. Simons Island, where she lives with a daughter, is an assortment of paintings by her talented son. Her favorite, which hangs over the mantel, is of Doug Sr. In a trophy case in one corner of the living room are the many silver loving cups and plaques accumulated by her husband in his brilliant but ill-starred

Doug Davis, Jr.

air racing career. A broad street along the north side of Atlanta's Hartsfield International airport bears the name, Doug Davis Drive. AQ

Mrs. Glenna Mae Davis of today, at home on St. Simons Island, GA

Wedell-Williams "44", in which Doug Davis lost his life during 1934 Thompson Air Races, Cleveland, Ohio. *(Line drawing by N. A. Ross)*

AVIATION QUARTERLY

ABOUT THE AUTHOR:
Clair Stebbins, a former president of the Aviation/Space Writers Association, wrote an aviation column for years and covered assignments all over the world for *The Columbus (Ohio) Dispatch*. He collaborated with General Paul W. Tibbets, pilot of the plane that dropped the A-bomb on Hiroshima, in the latter's autobiography, "The Tibbets Story." Stebbins' home is in Zanesville, Ohio.

(All photos by: Clair C. Stebbins)

Cessna C-145 "Airmaster"

In September 1938, the first new model rolled out designated as the C-39, however Cessna changed its designation to C-145, which reflected its power rating. Also, in June 1939, an Airmaster version with the new Warner 'Super Scarab' version of their 165 hp engine became available and Cessna Aircraft attained approval on October 3, 1939.

These type aircraft are compact high-winged cabin monoplanes with seating arrangements for four. The quickest recognition factor of these aircraft being lack of drag producing struts and braces, which were so common on other aircraft of that period.

The production of both C-145s and C-165s were intermixed as orders came in. The last of this series was rolled out in late 1941. Cessna Aircraft was already hip-deep in twin-engined "Bobcat" (T-50) production.

It was said you could fly the C-145 — under normal conditions — from New York to St. Louis, or from Chicago to Denver, and after paying for your fuel with a twenty-dollar bill you'd still have enough change to buy lunch for you and your party. AQ
-GES-

Reference: U. S. Civil Aircraft, ATC Numbers - 701 to 800, Volume 8. *(Aero Publishers)*

C-145 exterior and interior photos. Note close quarters for occupant seating. *(Photo: Sam Dick)*

AVIATION QUARTERLY

1946 Aeronca Champ: N85491

Owned by Raymond Murray of East Aurora, New York, this particular airplane — like so many — has seen many owners' during its four plus decades.

Manufactured by Aeronca of Middletown, Ohio and test flown by H. J. Rosing on August 8, 1946, with all log books and most repair information, from date of manufacturing, in the hands of Ray.

It's initial owner was Ernest Woods of Woods Flying Service, Port Richmond, Kentucky. Subsequent to the initial owner, eleven other aviation enthusiasts' purchased it prior to Murray. Most of N85491's base of operations was in Kentucky, Indiana and Ohio, prior to arriving in Western New York in 1976, and becoming the property of its current owner in June 1987.

"I have owned two Taylorcrafts, two Colts, an Ercoupe, and learned to fly in a Cessna 150. N85491 is my second Aeronca," Murray revealed.

Feeling that all aircraft have their good and bad points, he feels his favorite aircraft is the "Champ, except when a friend — owner of an 85 hp Piper Voyager — flies rings around me," he exclaimed!

Mr. Murray attended an Aeronca convention at the Middletown, Ohio factory in 1988, and discovered they were producing component parts under government contracts. "Probably making more money now than they ever did," Murray concluded. AQ

Roscoe Armstrong of Arlington, Texas

Pilot Roger Wolf Kahn flying the Burnelli "Guggenheim" monoplane, driven by two 95 h.p. American Cirrus III air cooled engines, which were mounted at the front corners of the center section.
(Photo: Courtesy of Robert E. Snowden, Jr.)

VINCENT J. BURNELLI
"Never A Member Of The Club"

by Kathryn Jones

Born: Temple, Texas, 1895.   Died: Long Island, N.Y., 1964.

Of all the 20th century's aircraft designers, none may be as unconventional, controversial or persevering as Vincent Justus Burnelli.

He was a prolific designer, yet only eight of his trademark "lifting body" designs ever advanced past the drawing board to actually fly. He was an innovator, holding more than 100 patents related to the lifting body wing, variable camber wing, landing gear, amphibian floats and other devices. Yet the aviation world never fully accepted his unique airfoil-shaped fuselage and branded him an eccentric.

He also was one of aviation's most tragic figures. While many aviators considered Burnelli a genius, his lack of business sense, disdain of politics and incredible run of bad luck produced a string of setbacks that Burnelli never was able to overcome.

Burnelli's idea was to make the dead weight between the plane's wings contribute up to 65 percent of its lift. He designed an airfoil-shaped fuselage, a modified "flying wing" that was squat, boxy and odd-looking. Even so, many of the top test pilots and military officers of the 1920s, 1930s and 1940s lauded Burnelli's funny-looking planes as the most efficient ever built. To this day, the lifting body concept remains his greatest legacy.

Twenty-five years after his death, a small but devoted band of pilots, retired military officers and aviation engineers are trying to revive interest in his lifting body design, believing its slower takeoff and landing speeds and structure offer significant safety improvements over conventional aircraft.

Military and commercial aircraft manufacturers and the federal government, however, have resisted building such a plane, saying it does not offer major advantages over conventional designs. Yet, Burnelli followers see his influence in lifting bodies such as the Space Shuttle and in flying wing aircraft such as the new B-2 Stealth bomber.

Leading the fight for the Burnelli design is Chalmers H. "Slick" Goodlin, the test pilot for Bell Aircraft who took the XS-1 on its first powered flight in 1946 and to the edge of Mach 1 before Chuck Yeager stepped in. Goodlin still runs the Miami-based Burnelli Co. in addition to his own air transport brokering business. He believes the reason Burnelli's designs aren't flying today is because of politics and greed.

"They've been sitting on this technology," he said. "It's because of politics and monopolism. We are not a capitalistic country, we're a monopolistic country. Innovations are unable to be funded because they aren't owned by the monopolists."

Burnelli's unconventional streak began in childhood. Born in Temple, Texas, in 1895, Burnelli studied the principles of aerodynamics by watching the buzzards that floated on the warm central Texas breezes. He built several airframes and covered them with feathers. They crashed, but he kept trying to improve his design.

Burnelli's father was an educated man of Italian heritage (Burnelli later shortened his name from the family spelling "Buranelli"). He wanted his son to have better schooling than was then available in Temple and sent Burnelli to Jersey City, N.J., to continue his education. There, family friends - the Moissant brothers - ignited his interest in aviation.

John D. Moissant had been the first pilot to fly a passenger across the English Channel. His brother Alfred had converted an old coffin factory into an aircraft plant and built small planes for exhibitions. Burnelli learned the basics of mechanical drawing from Alfred Moissant. He also participated in model meets of the New York Aero Club and began gliding, a feat that later earned him membership in the exclusive "Early Birds."

In 1913, at age 18, Burnelli became an aeronautical draftsman for Moissant. His job was to figure the aerodynamic drag on the new planes Moissant was building. He would total the amount of the components — the landing gear, struts, tail, fuselage and wing — and compare it with the lift provided by the wing.

Burnelli discovered the wing contributed almost all of an aircraft's lift, while the elements that produced drag consumed most of the engine's power. "I figured then that the airplane of the future would have a giant wing," Burnelli later recounted. "It looked like the quickest way to increase lift without exorbitant power. But that was in 1914 and I was sure the giant flying wing would probably come out in 2014. But I went ahead anyway and began designing just a segment of a flying wing."

Burnelli eventually rejected the idea of a pure flying wing, deciding it would have to be bigger than any airline or manufacturer would consider building. So he compromised his ideal for economics and designed a modified flying wing with the airfoil-shaped fuselage that became his landmark.

Early on in his career, though, Burnelli

Chalmers "Slick" Goodlin, at home in Miami, Florida. *(Photo: Compliments of The Burnelli Company.)*

concentrated on conventional aircraft. He designed his first plane in 1915 at Maspeth, Queens, with a friend, John Carisi. They demonstrated the open biplane at the old Hempstead Plains Aviation Field, later renamed Roosevelt Field. " "We used it for barnstorming,' ' Burnelli later recalled. "You could make $500 to $1,000 in those days working a fair, and that was big money."

A few years later Burnelli designed a night fighter, hoping it would fly in World War I. He wasn't successful in selling it to the military, but he did sell the plane to the New York Police Department for its aerial police operations.

After World War I, Burnelli became chief engineer of Lawson Aircraft Co. of Milwaukee and was in charge of designing and building the Lawson Airliner, the first enclosed cabin airliner flown in the United States. The plane set several records by successfully carrying 20 passengers on long cross-country flights and was hailed as the air transport of the future. But Burnelli later recalled he was disappointed with the airliner. It reminded him, he said, of a streetcar with wings. From that point on he pursued his lifting body designs.

In 1920 Burnelli teamed with T.T. Remington to build the first lifting-body plane, the RB-1, a 30-passenger plane. It had a wide, airfoil-shaped fuselage that housed two 55 h.p. Scottish-built Galloway Atlantic engines side by side. Judging from reports at the time, the RB-1 successfully demonstrated Burnelli's lifting body idea, but lacked good directional control. The first RB-1 was lost at Staten Island, N.Y., in 1923 in a violent storm.

V. J. Burnelli and RB-2 in 1925. *(Photo: Robert E. Snowden, Jr.)*

The RB-2 followed in 1924, credited as the first airfreighter. The plane, initially fitted with Galloway Atlantic engines, was demonstrated at Curtiss Field and at Mitchel Field for the Army Air Corps. By aircraft standards at the time, it was enormous: The RB-2 had an 18-foot-by-14-foot passenger cabin that could accommodate 25 passengers in "parlour car" comfort or carry 6,000 pounds of freight and a crew of three.

The RB-2 is probably best remembered, though, for the unusual load it carried in 1925. Hudson Motors used the plane to carry the Essex automobile, plus the full office equipment of an automobile salesroom, on an aerial sales tour. The plane reportedly was fitted with more powerful engines to transport such a heavy load.

The RB-2 was Burnelli's last lifting-body biplane. All of his subsequent designs were monoplanes. The CB-16, built in 1928 in Keyport, N.J., was the first multi-engine design with retractable landing gear. The aircraft was moved by barge to Newark, N.J., where Lt. Leigh Wade and Lt. James H. Doolittle took it up on a successful 40-minute maiden flight.

Frank Hannam, a retired aviation engineer who was a flight engineer for Burnelli in the late 1920s, said the CB-16 "far outclassed other airplanes" in those days. It was the first large transport with retractable landing gear and the first multi-engine corporate plane, since it was built for the P.W. Chapman Co. of New York and Chicago to transport its executives. The aircraft was equipped with an ice box and a galley.

But the plane had one big problem which

AVIATION QUARTERLY

Front view of a Burnelli CB-16, on the flight line at Newark Metropolitan Airport, circa late 1920s.
(Photo: Courtesy of Robert E. Snowden, Jr.)

Overhead, rearview of CB-16, probably taken at Newark Metro Airport.
(Photo: Courtesy of Robert E. Snowden, Jr.)

AVIATION QUARTERLY

Two different views of the UB-20, which probably reaped more publicity than all of the other Burnelli designs. *(Photos: Courtesy of Robert E. Snowden, Jr.)*

Hannam attributed to poor workmanship by the factory: The installation of the throttle and mixture controls were so close together on the plane that the throttle control jammed and Hannam was obliged to get behind the engines during landings and hold the controls separated. He was in that position when the CB-16 crashed at the Newark airport in 1929.

Hannam said he and Capt. Earl Stewart had flown out to meet the S.S. Leviathan, the largest steamship at the time, when a strong crosswind hit the plane. The 83-year-old Hannam, who now lives in California, vividly recalls what happened next: "We made about six attempts to land and finally sat down and skidded off the runway into the mud. I saw the left wheel go rolling off in the boondocks and the airplane came to a sudden stop, nose in the mud, bent props, busted landing gear, and some wing damage. I worked all night with the assistance of a crane driver and a crane on tracks that luckily was on the field building hangars. A crew came up from the factory and we patched up the airplane sufficiently to fly to the factory at Keyport.

After the CB-16, Burnelli began working on an even more unorthodox design for the 1929 Guggenheim Safe Aircraft Competition, designed to advance flying safety. Burnelli's entry was the GX-3, a bizarre-looking aircraft with full-span, high-lift flaps, chamber, deflected leading edges fitted with endplates and an unusual undercarriage.

In 1929, Burnelli teamed up with Inglis M. Uppercu to build the UB-20, the first American transport aircraft with smooth, stressed sky construction. The plane had a 90-foot span, was 52 feet long and stood 13 feet. It could carry 20 passengers in a 17-foot-by-12-foot cabin that featured large reclining chairs and berths in addition to a lavatory, a galley and baggage, mail and radio equipment compartments. The plane once carried a Ford car slung between the undercarriage legs for Sun Oil Co.

A similar aircraft, the UB-14 followed, powered by two 680 h.p. Pratt & Whitney Hornet engines mounted side-by-side in the center section leading edge. The famous round-the-world flyer Clyde Pangborn extensively tested the UB-14.

About the time the UB-14 was being designed, the Uppercu-Burnelli Aircraft Co. failed in 1932. The airplane actually was built by the Central Aircraft Corp., which Burnelli started in 1933 with a loan from the Reconstruction Finance Corp. Burnelli was able to acquire the plant and equipment of the Keyport factory where his previous airplanes had been constructed.

Burnelli claimed his lifting body designs protected passengers because the cabin section was protected by 65 percent of the total airframe structure compared with 15 percent for conventional designs. The UB-14 unwittingly became the ultimate safety test for Burnelli's design when it crashed near the Newark airport in 1935.

The report from test pilot Louis T. Reichers indicated he was flying at 195 miles per hour when the control system failed. Reichers flew the plane into the ground from about 200 feet altitude and estimated his speed at about 130 miles per hour when he crashed.

The plane cartwheeled, tearing off the engines and smashing the wings and tail section. But photographs of the spectacular accident show the plane's body stayed intact and no fuel leaked from the wing tanks. Photos of the cockpit and passenger cabin interior showed remarkably little damage. No one was injured and there was no fire, according to the report.

"It is my firm belief that the fact the box-body strength of this type combined with the engines forward and the landing gear retracted saved myself and the engineer crew and had the cabin been fully occupied and passengers with safety belts properly attached, no passengers would have been injured," Reichers wrote in his accident report.

"This crash landing, in my opinion, is an extraordinary example of the crash safety that can be provided by the lifting body type of design," said Reichers, who later became a top engineering official in the military transport command.

After the crash, a second prototype was built incorporating improvements in detail design, the UB-14B. Pangborn flew the plane nonstop from New York to England for demonstrations and the Scottish Aircraft & Engineering Co., of England became interested in building the plane under license. The company was later taken over by a receiver and its name changed to Cunliffe-Owen Aircraft Ltd., in 1938.

Cunliffe-Owen redesigned the UB-14 and designated it the OA-1, or the "Clyde

Burnelli UB-14 in Sept., 1945. Left to right: V. J. Burnelli with his chief pilot Clyde Pangborne and Mrs. Hazel Burnelli, wife of the innovator and inventor. *(Photo: Courtesy of Robert E. Snowden, Jr.)*

Burnelli and his CBY-3 in 1950. *(Photo: Courtesy of Robert E. Snowden, Jr.)*

The UB-14 was General Charles de Gaulle's personal air transport during World War II. The Clyde Clipper (UB-14) in 1939, was flown by Mollison to French Equatorial Africa.
(Photo: Robert Snowden, Jr.)

Clipper", while Burnelli was in England finalizing design negotiations over all assets, patents and the aircraft factory. Gen. Charles de Gaulle of France used the Clyde Clipper as his personal plane during World War II.

The last lifting-body plane built, the CBY-3 Loadmaster, is Burnelli's only surviving plane. Built by the Canadian Car & Foundry Co. in 1945 and equipped with two 1200 h.p. Pratt & Whitney R-1830 engines, it was designed to carry 22 passengers and a crew of three. The plane flew many demonstration flights and changed hands several times, before it returned to the United States, was re-engined and flown by the Burnelli Avionics Corp. It now sets in a field at the New England Air Museum in Connecticut, its restoration hampered by a lack of funds.

Professor F.K. Teichmann, who conducted some of the early wind-tunnel testing on the Burnelli lifting body concept at New York University, wrote of his approach: "All other things being equal, it is an acknowledged fact that a flying wing would be the most efficient design for a heavier-than-air craft. Such a design would be ideal, but by the time control and stability about the various axes have been considered, as well as the housing of powerplant, crew and cargo, various compromises with the ideal conception have to be made. Vincent Burnelli, in his designs, solved the problems in excellent fashion."

Burnelli tackled the problem of cargo space by devising a thickened mid-portion of the wing instead of a standard fuselage. Prof. Teichmann wrote that the thickened portion of the midwing is still a reasonably efficient airfoil, whereas the standard fuselage is not, "so that the next best thing to a flying wing is achieved."

Pilots who flew one of Burnelli's lifting body planes still speak of the experience with a touch of awe in their voices. Jay Brandt, a retired airline pilot and a stockholder in the Burnelli company, was flying DC-3s and Lockheed Constellations when he got the chance to take up the CBY-3 Loadmaster for a spin in 1948. "It was the greatest sensation," he recalled. "It was like going up in an elevator."

Brandt said the plane took off and landed in less than 500 feet. He estimated it was flying at 55 miles per hour when it touched down. "I had the greatest sensation of security," Brand said. "It took very little effort to fly it. I felt like I was driving a sportscar."

Friends and family members said that while Burnelli was an engineering dynamo, he was otherwise soft-spoken, unimposing, lacked business skills and judgement and didn't care about or understand politics. "He was the typical absent-minded professor," said Burnelli's nephew Dallas Swan, a real estate developer in Virginia who is the only family member still involved in the company. "He was a genius but he had no business sense. He got involved with a lot of people he shouldn't have."

Burnelli rebuilt his company after loosing control to the RFC, but from that point on he demanded absolute control. He closely guarded his patents and refused an invitation to join the Manufacturers Aircraft Association Inc., an exclusive and powerful organization of major aircraft manufacturers that had patent cross-licensing agreements. Burnelli's wife later said the decision made him an industry outcast.

What Burnelli thought was his big break came in 1941. He had won three government competitions to build planes for the U.S. Army Air Corps. The late Gen. Henry H. "Hap" Arnold, Chief of the U.S. Army Air Corps, during World War II, wrote in a letter to the Secretary of War that the Burnelli lifting fuselage project "should be carried through to its fullest experimental possibilities and probably to the ultimate conclusion for the purchase of a prototype."

"In my opinion, it is essential, in the interest of national defense, that this procurement be authorized," Arnold concluded.

Burnelli and his associates were invited to the White House Oval Office for President Roosevelt's signing of a directive authorizing purchase. As they waited for drinks to be mixed, FDR asked about Burnelli's financial backers. Burnelli named Joseph Newton Pew Jr. of Philadelphia, the chairman of Sun Oil Co., as one of his financiers.

According to written accounts by Burnelli's associates in the room, FDR flew into a rage, threw his pen across the floor and said he had helped finance the campaign of his 1940 Republican rival, Wendel Willkie. An Air Corps board of review soon after sealed Burnelli's fate.

In documents that were at the time stamped "Top Secret" and not released for public view for 20 years, the board said the Burnelli design was not new or novel and that the company didn't have the engineering staff or production capability to build the prototype "within a reasonable time." Interestingly, those same reasons are given today by aircraft manufacturers and the government for not pursuing the Burnelli lifting-body design.

The evaluation committee further recommended that the Air Corps inform Burnelli and "any other concern which may later possibly become interested in the Burnelli lifting fuselage that this design is of no interest to the Air Corps and for this reason no further correspondence, consultations or reviewing of data embodying this design ever again be considered by the Air Corps or the Materiel Division."

Vincent J. Burnelli, Aircraft designer.

Burnelli, devastated by the setback, worked on advanced jet designs from 1946 onward. But his efforts to find someone to build his planes were repeatedly turned down. Even European manufacturers wouldn't touch his designs.

At age 69, Burnelli died in Southampton, Long Island, following a stroke. The New York Times ran a long obituary, calling him a "pioneer in world aviation." In it, the Times quoted an aircraft designer as saying the reason Burnelli wasn't successful in popularizing his planes was because the aircraft industry kept developing higher powered engines that would fly their conventional designs.

Several air safety organizations, concerned about a recent string of air crashes, are renewing their efforts to promote the Burnelli plane as a safer, more "crashworthy" airliner than conventional aircraft. They cite the sturdy, boxy structure and the placement of engines and landing gear away from fuel tank supporting structures as safety features.

Goodlin also has been pressing the Pentagon, the Federal Aviation Administration and aircraft manufacturers to take another look at the Burnelli lifting-body. Goodlin hasn't made much progress convincing them, although the Burnelli Co. was invited to compete on a military program, the U.S. Air Force's proposed C-27 cargo plane.

Then last August, Goodlin received a letter from Deputy Undersecretary George P. Millburn that again brought up the 1941 board of review decision. "The matter has been explored thoroughly beginning with the Army Board of Review held in 1941 that found that the Burnelli concept provided no significant advantages for military aircraft," Millburn wrote. He closed the letter saying: "To our knowledge no issues remain unresolved — and we consider the matter closed."

Goodlin says the letter shows that, even after four decades, "the Burnelli blackball is still on." AQ

Burnelli Comments

by Bob Snowden

Capt. Bob Snowden, who contributed many of the previously unpublished photographs accompanying this article, never met Vincent J. Burnelli but feels he knows him well.

In Snowden's home in Cutchogue on New York's Long Island are Burnelli's personal papers, models, photographs and other artifacts that Burnelli's daughter Patricia gave Snowden. The retired airline pilot plans to use them for a Burnelli exhibition in an aviation museum the Long Island Early Flyers Club plans to build.

Burnelli was an early member of the Long Island Early Flyers Club. Snowden said the group plans to build the museum in Suffolk County and has been working on the project for three years.

Snowden, 75, began flying in 1928 and worked for the airlines for more than 30 years before retiring in 1980. He considers Burnelli a "genius" ahead of his time and has been trying to attract the government's interest in Burnelli's lifting body designs.

"When you go back 50 or 60 years, there really was not the opportunity for such a large airplane of that sort," Snowden said. "Everybody pooh-poohed it. It was a thing of the future at that time. It was the only one of its kind."

- KJ -

Burnelli CBY-3,
College Park, Maryland, 1957.
(Photo: Robert E. Snowden, Jr.)

CBY Loadmaster 194.
The sole remaining plane built
by Canadian Car and Foundry.
It was last flown in 1950.
(Photo: Robert E. Snowden, Jr.)

Burnelli Comments

by George E. Haddaway

George Haddaway, an aviation historian who knew Burnelli and saw one of the Uppercu-Burnelli planes fly, described Burnelli as a brilliant designer who was besieged with hard luck.

"All the big companies avoided him," Haddaway said. "He was a successful failure, a genius who never quite made the grade.

"He was a poor business man. It was his fundamental tragic trait. He usually received financial backing from the wrong people. He was an innovator and an inventor. His attitude was he didn't care about money and he sought it from the wrong sources."

Burnelli "had a complex," Haddaway said. "He thought everybody was against him. He was always a controversial figure and non-conformist. Aviation had been somewhat of a closed society up until World War II. Vince Burnelli was never a "member" of the club, President Roosevelt saw to that." AQ

— KJ —

Author

Kathryn Jones covers the aerospace and defense industry for The Dallas Morning News' business section. Prior to joining the News in 1986, she wrote about high technology and defense subjects for the Dallas Times Herald and before that was the Dallas bureau reporter for a Fairchild Publications computer newsweekly. Her interest in aviation dates back to her childhood in Corpus Christi, Tex., a "Navy town" where pilots trained and the Blue Angels precision flying team performed every summer. She would like to thank Dick Johnson, a retired LTV Corp. aerospace engineer, for sparking her interest in aviation history and introducing her to the work of Vincent J. Burnelli.

AVIATION QUARTERLY

DECEMBER 10, 1946 — MUROC ARMY AIR CORPS FLIGHT TEST CENTER, CALIFORNIA.

Bell Aircraft engineering test pilot, Chalmers H. "Slick" Goodlin stands before the Bell XS-1 after piloting the rocket ship on its first powered flight. Slick Goodlin made a total of 26 flights in the first two XS-1 aircraft before they were delivered to the U.S. Army Air Corps in June 1947.

AVIATION QUARTERLY

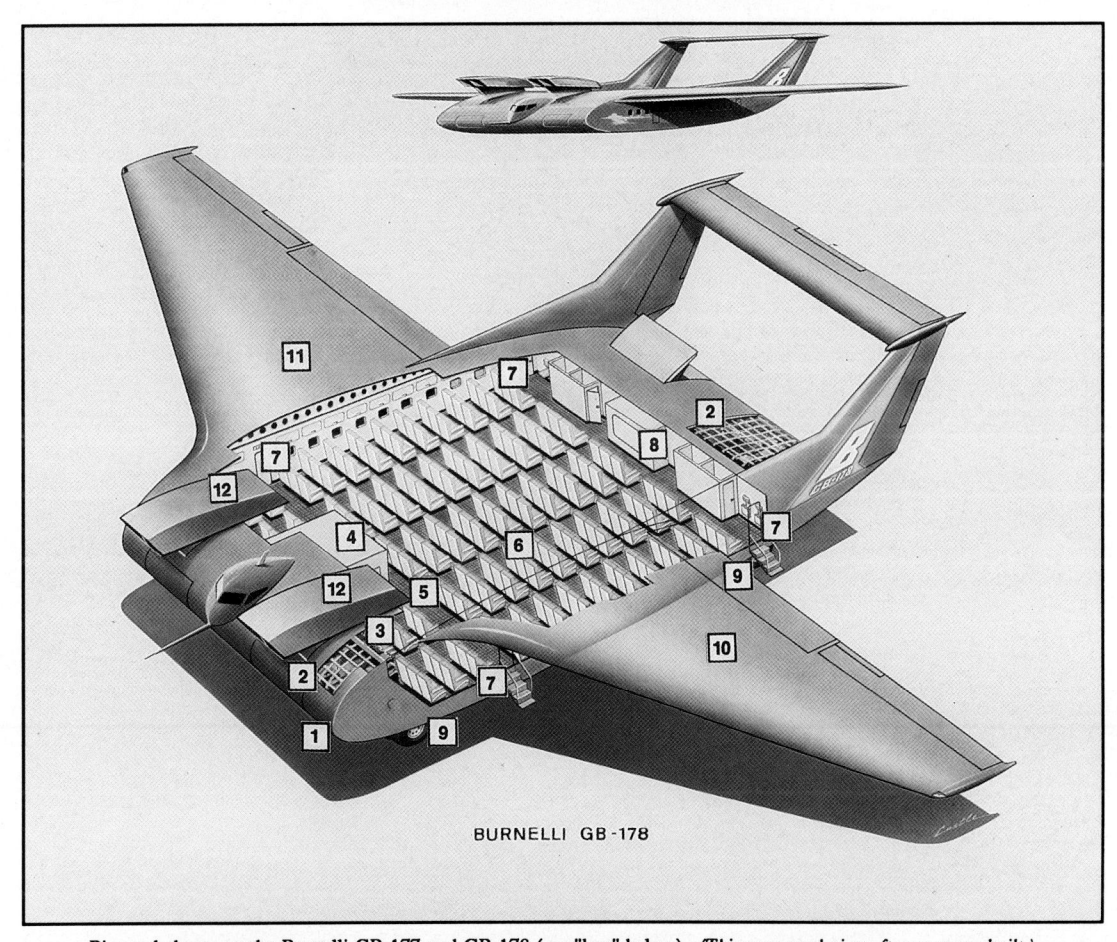

Pictured above are the Burnelli GB-177 and GB-178 (see "key" below). *(This proposed aircraft was never built.)*

KEY TO GB-177

1. Airfoil-shaped Burnelli fuselage eliminates parasite drag of tubular conventional designs while providing more "lift." It permits slower take-offs and lower landing speeds and surrounds passengers with 60 percent of aircraft structure which offers unparalleled "crashworthiness."
2. Baggage holds easily accessible for speedy loading/unloading.
3. Large storage racks for hand luggage.
4. Central large galley facilitates ease of service and provides secure equipment storage.
5. Greater floor area and cubic volume permit wider, more comfortable seats and wide aisles for easier access.
6. Rear-facing seats attached to main structure offer best crash protection.
7. Twin doors, both sides, permit fast embarkation/debarkation and emergency evacuation. Close proximity to ground eliminates need for expensive hi-loading, maintenance equipment, emergency slides, etc...
9. Quadricycle landing gear provides Rolls-Royce-like ground handling and braking qualities. Accessible in flight for emergency extension.
10. Fuel tanks in wings are isolated from engines and landing gear which virtually eliminates the crash fire hazards inherent in conventional airlines.
11. In emergency water landing, wings could be jettisoned, the fuselage becoming a boat.
12. Quickly detachable engine nacelle.
13. The Burnelli GB-178 dorsal-mounted "eyebrow" engines cause jet noise to be deflected skyward, substantially improving environmental operating conditions.

AVIATION QUARTERLY

BURNELLI GB-888

AVIATION QUARTERLY

Burnelli RB-2 circa 1925. *(Line drawing: N. A. Ross)*

AVIATION QUARTERLY

NC 868N
CURTISS FLEDGLING
The Thrill Of Discovery

by Dick Hopkins

Tim Talent, an airplane restorer from Springfield, Oregon had told me that a man near Medford, Oregon owned a 1929 Curtiss Fledgling. I finally discovered the owner's name and location through Boardman C. Reed, who owned a Tim Collegiate parasol from 1929. His aircraft is very similar to the Collegiate NC 888E, I worked on in 1946.

I called the number I had been given to talk to Mr. Burrill, and his brother Glenn answered. He invited my wife and I over to the Burrill Lumber Company's small airstrip. Upon arrival he immediately rolled out the big old biplane, with my assistance. I couldn't help but be impressed with its large size and the beautiful restoration job the Burrill brothers had done. The discovery was well worth the trip and made me realize how fast forty-two years had gone by since I'd seen my last Fledgling.

We returned to Oregon and once again feasted our eyes and emotions upon NC 868N, on a beautiful March 17, 1989 spring day.

While taking photos of this 60 years old biplane, with its 40 feet wing span, I realized it is probably the last double bay biplane manufactured in the United States.

Editor's Note:
In early fall of 1988, I had a telephone conversation with Dick Hopkins of California. Within a short time he mailed me a few pictures of a Curtiss Fledgling, owned by another aviation enthusiast in Oregon, Gene Burrill.
Realizing the near extinct status of this aircraft model, Mr. Hopkins made another trip to Oregon and came away with some outstanding photos seen with his story.   -GES -

Curtiss NC 868N Fledgling owned by Gene Burrill of Prospect, Oregon. *(Photo: Dick Hopkins)*

Above: Shadows of an evening sun on NC 868N. *(Photo: Dick Hopkins)*

Owner Gene Burrill and his Fledgling. *(Photo: Dick Hopkins)*

The double bay meaning it has two sets of interplane struts between the wings, instead of the single one near the end of each wing tip.

Also, the elevator control cables are on the exterior of the fuselage beginning at the control horn located in the gap between the lower wing and the fuselage. The cables extend back into two exterior pulleys located at the top and bottom of the stabilizer leading edge. Even for 1929, this seems a little archaic yet so unique. The unusual curve of the balanced rudder and elevators, the eight rubber biscuit shock absorbers on the main landing gear struts, all add to the fact that this is truly an unusual airplane, yet with continuing traces of the WWI biplane.

When Gene Burrill of Prospect, Oregon purchased NC 868N (or what was left of it), in 1969, from J.D. (Red) Berry who, at that time, resided in Fairbanks, Alaska, Burrill trucked the Fledgling back to Oregon with some Alaskan temperatures reaching 45 degrees below freezing. The aircraft sat for 10 years in Oregon, before Gene and brother Glenn began restoration. Gene stripped the aircraft fuselage down to barebones and started a meticulous restoration, involving research and many hours of hard, careful work, which culminated in its first flight in 1982.

This Curtiss Fledgling is all original with the exception being N3N wheels, with shoe brakes and a tail wheel replacing the tail-skid for better ground control. The Fledgling's shock absorber is eight rings, ("biskets") built into the landing gear main struts. Burrill has the original wheels hanging from the hangar rafters — next to

Close up of exterior rigging of Fledgling cables and pulleys on vertical and horizontal stabilizer. *(Photo: Dick Hopkins)*

A good illustration of 'biskit' type shock absorbers on Curtiss NC 868N. Note eight rings in all. *(Photo: Dick Hopkins)*

a pair of 10" x 40" Goodyear spoked wheels — reportedly used by Admiral Byrd on his Fokker trimotor.

NC 868N is the only flying Curtiss Fledgling with the original Curtiss Challenger six cylinder engine, with the hand forged Curtiss Reed prop.

This rare double row three cylinder radial is the 185 hp version, because the Burrills' installed a modified cam kit, which could have been readily purchased in 1931.

To my knowledge, the only other operational Curtiss Fledgling is at the Old Rhinebeck Aerodrome in Upper New York State.

One of the other two Fledglings is N2C-2 Trainer A-8529, at the Naval Aviation Museum, Pensacola, Florida. The fourth Curtiss Fledgling is 263 H, believed to be in a museum somewhere in Brazil. AQ

About Dick Hopkins

Regarding my interest in Aviation, my father influenced me. He was raised in Southern California during the era of experimental flight.

As a 13 year old, he attended an air meet at Domingez Field in 1910, and saw Roy Knabenshue fly his cigar shaped dirgible. He controlled the airship's ascent or decent by walking toward the front or rear of the open-frame-catwalk slung beneath. My Dad also made a model of the Farman after the meet.

In the summer of 1945, at the Eugene, Oregon Airport I was employed as a mechanic's helper. This was my first exposure to the Timm Collegiate NC888E built in May 1929. I also soloed in an Aeronca Defender, (converted WWII liason plane) in 1945.

I am an architect, and also a member of the American Aviation Historical Society. I also collect old electric trains and naturally — old toy airplanes.

Last of its kind, the Curtiss Challenger six cylinder radial engine.
(Photo: Dick Hopkins)

AVIATION QUARTERLY

Pull the chocks, she's ready to go.
(Photo: Dick Hopkins)

Looking down on the cockpit areas of NC 868N.
(Photo: Dick Hopkins)

AVIATION QUARTERLY

NC 868N and another Fledgling out of the Old Rheinbeck Aerodrome in up-state New York, continue to be certified airworthy.
(Photo: Dick Hopkins)

Burrill's Curtiss Fledgling front view.
(Photo: Dick Hopkins)

AVIATION QUARTERLY

NC 868N vertical and horizontal stabilizer with external control cables and pulleys. *(Photo: Dick Hopkins)*

AVIATION QUARTERLY

Runups

AVIATION QUARTERLY

(AQ collection)

Back the attack - with War Bonds

MAKERS OF THE UNITED STATES AIR FORCE

7

William E. Kepner: All the Way To Berlin

Paul F. Henry

Maj. William E. Kepner

William Ellsworth Kepner was always a scrapper. He was the kind of American military man *Time* magazine's World War II reporters loved to write about. He was a general who, at the age of fifty, flew fighter missions over German-occupied territory—a tough, laconic veteran who led by example. But Bill Kepner was more than that. He had been a pioneer in the Air Corps' brief flirtation with balloons and airships, an early explorer of the stratosphere, and a defender of fighter aviation in the years when the bomber was king. His tactical innovations as head of VIII Fighter Command during World War II were a lasting contribution to the development of air warfare. They played a major part in defeating the Luftwaffe and assuring success of the Combined Bomber Offensive that destroyed the military infrastructure of Nazi Germany.

George Harvey Kepner and his wife Julia Ann had given their son William, born in predominantly rural Miami, Indiana, in 1893, a solid foundation of Midwestern values. Young Bill demonstrated old-fashioned stubbornness and a fierce streak of independence by leaving Kokomo High School one marrow-chilling November day in 1909 to enlist in the United States Marine Corps. Thus a promising Kokomo sophomore began a relationship with the armed forces that was to occupy nearly forty-one years of his life.

Marine Corps ways agreed with the athletic sixteen--year-old Kepner who, thirty years later, reflected that the Corps "isn't a particularly easy way to live, but it is a very satisfactory way." Convinced, nevertheless, that he needed to complete his education, Kepner accepted an honorable discharge and Marine Corps Good Conduct Medal in November 1913, and returned to school with plans for a medical career. This dream was short-lived, however. In 1916, Indiana National Guard units were being called up for service on the Mexican border; Bill Kepner applied for and received an officer's commission. He accompanied his unit to Mexico, was augmented into the Regular Army and promoted to first lieutenant on June 14, 1917.

Kepner, by summer's end a captain, went overseas with the 4th Infantry, 3d Division of the American Expeditionary Force. As "I" Company commander, he saw action in the Chateau-Thierry and St. Mihiel offensives and was decorated for individual heroism in hand-to-hand fighting. Combat brought out the natural aggressiveness and determination that grew out of Bill Kepner's Marine training and flinty personality. "The only

ALL THE WAY TO BERLIN

time you can quit with any self-respect," he said, "is when you are dead."

By 1918, Kepner had developed a passionate interest in flying. That interest grew, he wrote, "especially after using my Infantry Company's ground fire to drive off three German fighter planes who forced a French pilot to land in our Company area across the Marne River at Chateau-Thierry." His article, "Reminiscences of an LTA Pilot," which appeared in the September 1978 issue of *Air Force Magazine*, recalled an early try at transferring to flying duty:

> After Chateau-Thierry and St. Mihiel, I asked, at an officer's meeting, to transfer to the Air Service for airplane pilot training. Colonel Halsted Dorey replied "Yes, if you want to be a frill. You have a man's job where you are." When the meeting was over, he put his arm around my shoulder and promised me a battalion....

Kepner was given the 3d Battalion for the Meuse-Argonne campaign. Under his command the unit captured enemy strong points at Farm de Madelaine and Mt. Faucon. These were key German positions which, once lost, helped accelerate the final surrender. This battle also brought Kepner's World War I combat to a close; he was seriously wounded and spent months recuperating in a French hospital.

Some fourteen months after the Armistice, Captain Kepner was back in the States, assigned to the 61st Infantry at Fort Gordon, Georgia. He immediately applied for pilot training, preferably at Arcadia, Florida, an airplane station. He got Arcadia, all right, but it was in California where all the U.S. balloon schools had been combined at Ross Field.

Most Army officers considered balloons to be in somewhat the same category as pack mules—cantankerous and unpredictable beasts of limited utility. Besides, Kepner had observed that during the recent war, "It seemed as though, sooner or later, every balloon was shot down." He wired an urgent message to the Army's Adjutant General pleading that a mistake had been made in his assignment. The sharp reply from headquarters read: "There is no mistake. Go to Arcadia, California, and no more direct contacts out of channels." He reported to the school in November 1920 and graduated with the rating of Balloon Observer the following May. Another student who completed the course that year was Oscar "Tubby" Westover, who later became Chief of the Air Corps.

Kepner's reluctant acceptance of his professional fate did not obscure the more positive aspects of a lighter-than-air career. Some of the Air Service's best-known pilots, like Frank Lahm and Benjamin Foulois, had begun their flying careers in balloons. Sport ballooning in the 1920s was an international pastime, and the Army had traditionally participated in racing events with teams selected from among its crack balloon crews. Though Kepner could not have anticipated it, the Air Service was to embark in the 1920s on a short-lived airship program. It was as a sport balloonist and airship pilot that Bill Kepner achieved recognition and began earning his place as an aviation pioneer.

Ballooning was by its very nature a demanding activity. Floating about under a creaky fabric bag of highly flammable gas, at the mercy of fitful winds, and standing in the 1,500-pound basket, which acted like a berserk pendulum in the slightest turbulence,

Below: WWI balloon on observation duty. *(AQ Photo)*

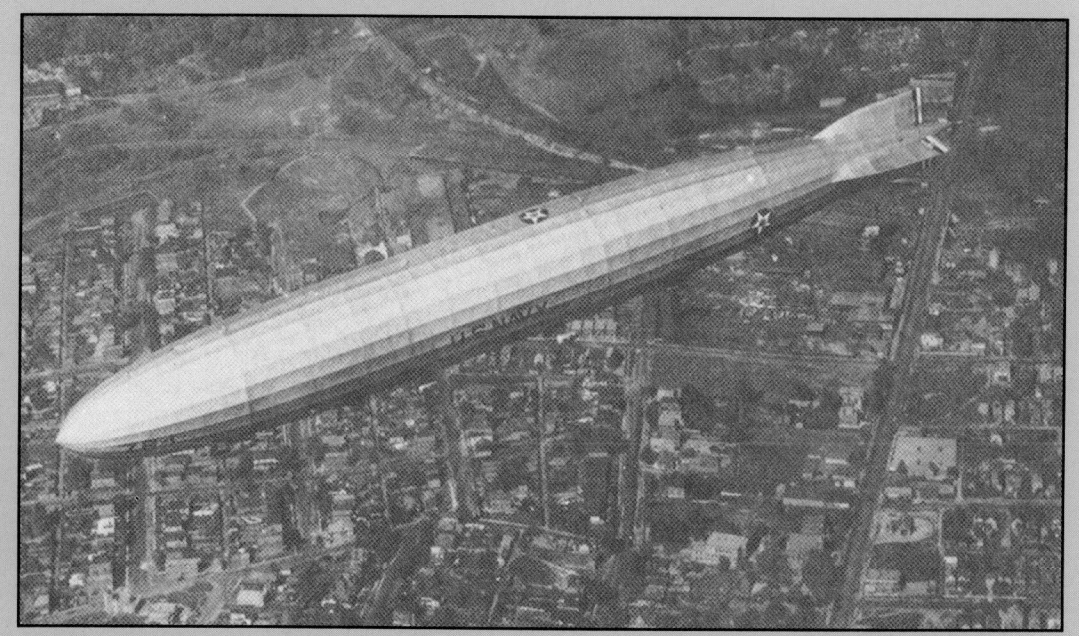

The ZR-1 *Shenandoah* was completed in September 1923; it was lost two years later in a storm over Ohio. *(AQ Collection)*

MAKERS OF THE UNITED STATES AIR FORCE

A U. S. Navy balloon of the mid-twenties, similar to those of the Air Service.
(Photo: History of Aviation Collection, UTD, Texas.)

ALL THE WAY TO BERLIN

required at the least a strong constitution. Balloonists of the day acted as their own meteorologist, logisticians, navigators, repairmen, and general all-around roustabouts. These hardy aviators required thorough preparation in all the basic skills because, Kepner reasoned, "only then can they expect to have any idea where they are apt to go" once that first bag of ballast goes overboard.

Following graduation from the balloon school, Captain Kepner was named Commander of the 32d Balloon Company, but his military experience in balloons was destined to be short. These floating observation posts were being replaced by large, engine-driven airships. The Army was systematically deactivating all balloon units. So, after only seven months in command positions with two different balloon companies, Kepner was sent to the Army Airship School at Langley Field, Virginia. Though a student, he also was named Commander of the Airship School Detachment. There was a lot going on in lighter-than-air. The Army and Navy were vigorously exploring the military utility of airships, and young pilots enjoyed the pleasant prospect of flying non-rigid, semi-rigid, rigid, and pressure rigid airship types. To fledgling airship pilots like Bill Kepner, "military lighter-than-air (LTA) looked like a serious business that was well started."

To Kepner fell the additional privilege, while still a student, of serving aboard the huge semi-rigid airship *Roma* during its initial familiarization flights. The *Roma*, acquired from the Italian government, crashed during a test flight in February 1922, and all but eleven of the forty-five crew members aboard were killed. Kepner had been ordered at the last moment to move a small airship out of *Roma's* way so she could be

Kepner, while a student, had the privilege of serving aboard the ill-fated airship *Roma* during its initial familiarization flights.
(Photo: History of Aviation Collection, UTD, Texas.)

MAKERS OF THE UNITED STATES AIR FORCE

taken from the hangar. Minutes later, *Roma* crashed in flames as Kepner watched, stunned, from his own airship.

Kepner graduated from the Airship School in June 1922 and left Langley in the heat of the summer to command the 18th Airship Company at Aberdeen, Maryland. His new station was home of the Ordnance Proving Ground, and it was there that his flight-test experience began with bombing tests and long-range navigation sorties in Army C-2-class airships.

* * * * *

Bigger things were in store for Kepner as spring returned to Maryland in 1923. In March, he was selected to attend rigid airship training with the Navy at Lakehurst, New Jersey. In exchange for supplies of helium gas, the Navy offered to provide Air Service crews with rigid airship training. This was a unique chance for Army people to master the largest active airships in the world, and to prepare for possible joint operations with the Navy. Kepner graduated as a Naval Aviator, Zeppelin Pilot, and served as assistant navigator on the *Los Angeles*, a rigid airship.

By 1926, when Bill Kepner finished his tour of duty at Lakehurst, Scott Field near St. Louis had become the Air Service's airship training center. The Airship School also flight tested new airship designs. In April, about three months before the Air Service became the Air Corps, Captain Kepner reported for duty at Scott. With some 340 hours of flying experience at Lakehurst, he was the logical choice to test-fly the RS-1, newest and largest semi-rigid airship in existence. This mammoth vehicle had a 700,000-cubic-foot gas capacity, two engines with 17 foot propellers, and a top speed of 75 miles an hour. The RS-1 made a number of cross-country flights and participated in Army combat maneuvers near San Antonio, Texas. Kepner, now chief of the RS-1 test program and assistant commandant of the Scott airship training center, personally assayed the durability of the craft while flying through towering midcontinental thunderheads between Vicksburg and Memphis on the return trip to Scott Field. "Several times the airship was sucked up into the clouds then forced down to tree-top level," Kepner remembered. "The nose frame was crushed and the helium gas containers developed leaks." Using blankets and hastily improvised patching materials to stop the leaks, Kepner and crew brought the RS-1 home from what turned out to be its last long voyage. This proud but scarred airship was dismantled in the winter of 1928.

Kepner also found time in 1927 for some "outside activities." He and Lt. William Eareackson joined two other Air Corps teams in the National Balloon Race at Akron, Ohio. They earned a third place after drifting all the way to the Maine coast and landing at night in a fog-enshrouded graveyard. A short distance further and they would have been out to sea. This showing earned Kepner and Eareackson a spot on the U.S. balloon team and a chance at the 1927 International Balloon Race. They finished in the middle of a field of fifteen contestants after a harrowing experience. Plagued by bad weather from the start, Kepner and his crewman were swept up in a thunderstorm to an altitude of 27,000 feet without benefit of oxygen.

In the 1928 National Race, Kepner and Eareackson led the entire field, though not without cost. Once again they were trapped within the black folds of a thunderstorm. Even after the balloon at last cleared the clouds, the wild ride was far from over. Years later General Kepner still vividly remembered that ride:

> We went at great speed down a valley, took out three 20,000 volt electric lines, and crashed into a six-line assembly. Our wet drag rope crossed two of the lines and put out all electricity in the area. We then hit a two-arm railway telegraph pole and hung there until we could push off and go on throughout the night to win the race, landing the next morning at Weems, Virginia, just before going out to sea.

During the storm, the two airmen had attached their parachute release handles to the balloon's rope rigging and sat on the edge of the basket awaiting their fate. Three other pilots taking part in the race were either burned or killed by lightning.

Their victory earned Kepner and Eareackson the right to represent the United States in the 1928 Gordon Bennett International Balloon Race. This time they had relatively clear sailing compared to the drama of the National Race. They took first place by a wide margin and brought home the King Albert of Belgium Trophy—the first Air Corps team to win it since Lt. Frank Lahm in 1906.

In March 1929, Captain Kepner became Chief of the Lighter-Than-Air Branch, Experimental Engineering Section, at Wright Field, Ohio. He spent over three months during the summer test-flying a metalclad airship, the ZMC-2. Kepner believed this new design "was the strongest-type airship ever built," and "offered to fly it through a

ALL THE WAY TO BERLIN

U. S. Army Semi-Rigid Airship of the early to mid-twenties with crew gondola underneath. *(Photo: History of Aviation Collection, UTD, Texas.)*

line squall to prove it." It was becoming increasingly clear, however, that the Air Corps did not share his confidence. Many airships were being "retired" or passed off to the Navy. Kepner himself began casting about for a way to realize his earlier dream of being an airplane pilot. He had done just about all there was to do in lighter-than-air. His balloon-race victories and experimental airship testing invested him with a well-deserved reputation. The King Albert of Belgium Trophy, with Kepner's name prominently displayed, decorated the Air Corps Chief's office. It was a good time to move on.

While on detached duty for flight tests at his naval airship alma mater at Lakehurst, Kepner learned of his assignment to the Air Corps Primary Flying School at March Field, California. He completed training there in October 1931, the same month he was promoted to major, and graduated from the advanced course at Kelly Field, Texas, in February 1932—a little more than a month after his thirty-ninth birthday. At last, he pinned on the coveted airplane pilot wings that had eluded him for some twelve years. His balloon, airship, and aircraft ratings—six in all—put him in a class of experience few airmen had achieved.

The vagaries of the service being what they were, the Air Corps (with no little irony in Kepner's view) promptly returned him to Wright Field for more airship duty. This time, the craft was Goodyear's TC-13. After two months of flight test, Kepner *finally* bade experimental airships farewell forever, and, in March 1934, was installed in the Air Corps Materiel Division at Wright Field as Chief of the Purchase Branch. He had not seen the last of lighter-than-air piloting, however. Bill Kepner said a final good-bye to balloons with characteristic dash—in a stratospheric balloon flight for the National Geographic Society.

* * * *

This flight was a national event that captured public attention in the summer of 1934. It grew out of earlier ascents by Capt. Hawthorne C. Gray in which Kepner had assisted, both in laboratory experiments and as a ground crew team member. On the second of Gray's flights in his open-basket balloon he reached 42,470 feet but died in the

MAKERS OF THE UNITED STATES AIR FORCE

attempt when he inadvertently dropped a full oxygen bottle as ballast instead of an empty one.

Kepner's 1934 mission, under the joint auspices of the National Geographic Society and the Army Air Corps, was far more sophisticated than Gray's fatal 1927 try. The enclosed gondola, which contained more than a ton of scientific equipment and 7,000 pounds of ballast, was constructed of a lightweight magnesium alloy. Christened *Explorer I*, it was attached to a balloon more than 300 feet high with a three-million-cubic-foot gas capacity. Two other crew members, Capt. Albert W. Stevens and Lt. Orvil A. Anderson, accompanied Kepner to perform scientific experiments and help control the balloon in flight. The purpose of this exploration, as the *National Geographic* rather dramatically put it, was in part to investigate "... the mysterious ozone layer of the upper air." More specifically, the scientific objectives were to analyze cosmic radiation, define the ozone layer's position, check air composition, and record accurate pressure-temperature-altitude data.

In specifying a launch site, Kepner laid on some exacting criteria. "I need," he said, "a hole 400 feet deep with vertical walls; a 500-foot-square grassy meadow at the bottom, with a 20,000 volt electrical line; a railroad and a first-class truck highway running through it; and, if possible, I would like a good trout stream running through it." As luck would have it, the Chamber of Commerce in Rapid City, South Dakota, provided just such a place—trout stream and all. Kepner and his crew dubbed it the "Stratobowl" and made preparations for a late July launch.

The lift-off occurred on July 27, 1934, in something of a circus atmosphere. Some 120 cavalry troops, watched by Sioux Indians in full ceremonial regalia, nervously displayed their lack of experience as balloon handlers, while hundreds of local residents cheered and felt-hatted newsmen spoke busily into large microphones. The ascent was uneventful until about 57,000 feet. There, a large hole appeared in the balloon bag and the crew was forced to check the ascent. They reached 60,613 feet before Kepner was able to arrest the climb and begin a controlled descent. On the way down, the balloon fabric deteriorated increasingly and, after depressurizing the gondola at 20,000 feet, the tree men put on parachutes in case they had to abandon ship. In the meantime, a national radio audience (and the Air Corps Assistant Chief, fellow balloonist Oscar Westover) listened in on the drama. At about 4,000 feet, the balloon, now full of oxygen-contaminated hydrogen, exploded and the crew bailed out, Kepner going last at about 500 feet.

All three members of the expedition landed safely in freshly tilled soil near Loomis, Nebraska, and despite the gondola's destruction, Kepner estimated that the flight had been 60 to 70 percent successful. With this rather wooly adventure, Bill Kepner ended his lighter-than-air career. It had been a fitting glorious culmination of air achievements that began for him at the Army Balloon School in 1920. The *Explorer I* mission was a ground-breaker for later stratospheric experiments crucial to the modern Air Force. Ironically, it occurred the same year that the Air Corps officially washed its hands of lighter-than-air aviation. Kepner, too, had new interests. The balloon and the airship were now twin anachronisms, bulbous dinosaurs of aviation. Bill Kepner and the Air Corps left them behind at the same time.

* * * * *

Crew members of the flight from left to right: Maj. Kepner, Lt. Orvil Anderson, and Capt. Albert W. Stevens. *(USAF Photo)*

ALL THE WAY TO BERLIN

Major Kepner's partners in the flight of *Explorer I*, Captain Stevens and Lieutenant Anderson, had urged him to join them in a second stratospheric mission scheduled for the summer of 1935. The prospect of breaking the 1934 Russian altitude record, which still stood following the aborted *Explorer I* attempt, strongly appealed to Kepner, but he felt compelled to say no. He had postponed attending the Air Corps Tactical School in order to pilot *Explorer I*. Now Maj. Gen. Douglas MacArthur, the Army Chief of Staff, paid Kepner the compliment of asking the Air Corps Chief to inquire whether Kepner would prefer to command the second stratosphere flight or go to school. Kepner chose the latter and reported to the Tactical School, Maxwell Field, Alabama, in the summer of 1935. He was in good company: class members included Ira Eaker, Benjamin Chidlaw, and Nathan Twining, among other future Air Force giants.

Kepner, by virtue of seniority, was class president. He and his seventy fellow students faced an intensive nine-month course designed not merely to teach air strategy and tactics, but to prepare career officers for staff responsibilities. Among the Tactical School faculty members were some of the Air Corps' best thinkers. Men like Harold George, Haywood Hansell, Laurence Kuter, and Claire Chennault debated the issues of air power in what was a vigorous, sometimes acrimonious, academic forum. The curriculum was strongly influenced by the War Department General Staff; only about half of it was allocated to air subjects, but school authorities allowed lecturers and students to discuss their ideas on the use of air power, free of restraint. The willingness of Tactical School theorists to hear opposing views regarding air power employment may have been genuine, but there is little doubt that in the mid-thirties, bomber advocates, who believed that strategic bombing alone could win the next war, held sway over a small band of pursuit enthusiasts led by Claire Chennault.

Kepner was urged privately by Captain Chennault to champion pursuit, yet his interest in the argument sees to have been in practical hardware rather than theoretical doctrine. But the conflict between Tactical School bomber and pursuit advocates was fundamentally doctrinal and consequently all the more extreme. The bomber theorists believed fervently that heavy bombers flying in formation could penetrate enemy defenses during daylight hours in order to bomb with precision, and do so at an acceptable cost. That error in judgment extracted an extremely high price in bomber losses during the early months of U.S. participation in World War II. It presaged the long-range fighter escort problem which lay in Kepner's future, waiting to challenge him in the European theater, long after he had left the theorists at Maxwell to their heady dispute.

Before leaving Alabama, "Kill" Kepner became part of another pioneering venture. His friend and classmate, Ira Eaker, had conceived the idea of making a flight from coast to coast guided only by aircraft instruments. Its purpose was to demonstrate the effectiveness of extended navigation using radio and instrument aids. Eaker asked Kepner to pilot an observation aircraft flying alongside to reduce the hazard for Eaker of flying "blind" under an instrument hood. No stranger to experiments, Kepner readily agreed. Toward the end of the Tactical School course, the two aviators tried the idea out on the Commandant, Col. John F. Curry.

Explorer I before flight

With fabric torn, *Explorer I* falls to earth.

Spring 1989 • AVIATION QUARTERLY • 185

MAKERS OF THE UNITED STATES AIR FORCE

He encouraged Eaker to seek permission from the Air Corps. In May, the flight was approved and on June 2, 1936, Eaker and Kepner took off from Mitchel Field, Long Island.

They made the journey in two Boeing P-12s, an aircraft of proven reliability that had been a first-line pursuit plane since its addition to the Air Corps inventory in 1928. The trip, made in hops because of the P-12's limited range, took four days, with Eaker going under the hood as soon as his P-12 was airborne and remaining there until they were over the next landing field. Bad weather along part of the route forced Kepner to fly close formation on Eaker, relying solely on the "blind" P-12 for flight attitude and navigation.

Kepner was promoted to lieutenant colonel on June 16 and assigned as a student at the Command and General Staff School, Fort Leavenworth, Kansas. The years between 1937, when he graduated from the Staff School, and 1941 found Kepner in a variety of positions. As the nation moved somnolently but inevitably down the road to another world war, he was getting himself and the air arm ready to fight. Following the course at Leavenworth, he spent a year on the General Headquarters Air Force staff at Langley Field, Virginia, before assuming command of the 8th Pursuit Group at the same field. This organization and its sister pursuit group (the 1st at Selfridge Field, Michigan), were equipped with a mix of P-12s, Boeing P-26s, Seversky P-35s, and Curtiss P-36s. The P-26, an open-cockpit aircraft with fixed landing gear, was ancient technology by aviation standards, while the P-35 had proven hopelessly underpowered at altitude. Although the P-36 was vastly superior in performance to other first-line pursuit aircraft, it was available in only limited numbers. In the fall of 1938, Kepner was selected to command all defense aviation during the Fort Bragg maneuvers. This comprehensive joint exercise pitting pursuits against the new B-17 bomber was used with great success by Maj. Gen. Frank Andrews, the GHQ Air Force Commander, to assess fighter and bomber capabilities for the Chief of the Air Corps.

Kepner's tour at Langley concluded with his promotion to colonel and assignment to Mitchel Field as executive officer of the 2d Wing in February 1940. He shortly moved up to executive officer for the Air Defense Command, and when the First Air Force was organized in January 1941, Kepner became its first chief of staff. The following August, he organized and commanded the First Air

Left: Boeing P-12 of the U.S. Army Air Corps, circa 1928 to mid-late 30s.
(Photo: History of Aviation Collection, UTD, Texas.)

ALL THE WAY TO BERLIN

Support Command and during the Carolina maneuvers that fall, commanded all aviation under First Army. The Air Corps, as in the Fort Bragg exercises of 1938, was again flexing its muscles in preparation for the war many felt was just around the corner. Indeed, the Carolina maneuvers were to be the last "practice." After December 7, 1941, the game was for real.

* * * * *

Christmas 1941 was a dismal holiday for the American people. Customary good cheer had vanished in the aftershock of our sudden embroilment in the war. We clearly had been caught unprepared at Pearl Harbor. Our feeble resistance on Oahu seemed a sign of military weakness. During the years since Colonel Kepner had commanded the 8th Pursuit Group in 1938, however, a great deal of progress had been made in strengthening American air power. Since mid-October of that year, President Roosevelt had been emphasizing the need to increase U.S. air defenses. In 1939, Charles Lindbergh toured Air Corps bases examining equipment and visited most of the country's major aircraft producers. He reported his findings to an Air Corps board whose purpose was to devise a plan for aircraft and weapons development. Though it would be difficult to substantiate direct benefits of the so-called "Kilner Board," the period 1939-41 did become a time of frantic aircraft industry activity, much of it to support Great Britain and France, which was not exceeded until the months after Pearl Harbor.

The first B-17s had been delivered to Langley Field in March 1937, and the Air Corps, despite the usual fiscal and political obstacles, continued to buy them in slowly growing numbers. Pursuit aviation also had seen some changes, as Kepner discovered when he took over the IV Interceptor Command two months after the war started. The Bell P-39, Republic P-47, and Lockheed P-38 had replaced the older, slower, and less capable machines of Kepner's Langley Field days. Ironically, the Air Corps Tactical School arguments of the 1930s about pursuit versus bomber employment had persisted, and now (in 1942) the bomber people still were very much in the doctrinal driver's seat. Bomber preeminence had not been materially affected even by the air lessons of Spain's civil war and the Battle of Britain, which clearly showed the limited effect of strategic bombing on civilian morale and the essentiality of fighter escort to keep bomber losses at an acceptable level.

Kepner, a spanking new brigadier general in February 1942, tackled his duties at IV Interceptor Command (later IV Fighter Command) with characteristic verve and optimism. As he once said of himself, "I am much happier when the going is difficult than I would be if it were all calm and rosy." Faced with a contingent of inexperienced flyers and new, untried equipment, Kepner laid out a course of intensive training. He radiated confidence in America's ability to wage and win the war.

Kepner's deep belief in tactical flexibility found its natural expression in the fast-maneuvering, powerful fighter planes under his command. He was eager to join the units already in combat. "How I wish I could be with them," he had written to Larry Bell,

Above: Boeing P-26 with fixed landing gear. (*Photo: History of Aviation Collection, UTD, Texas.*)

MAKERS OF THE UNITED STATES AIR FORCE

An unknown Artist's rendering of pilot and P-39s. *(Look Magazine, 1942)*

president of Bell Aircraft. But nearly a year passed before General Arnold, in the spring of 1943, selected him to run the VIII Fighter Command, European Theater of Operations. Kepner had done a superb job organizing and training the IV Fighter Command; Arnold had a problem that needed solving, and Kepner was the right man to solve it.

The bomber commanders of Eighth Air Force, despite early optimism about autonomous bomber operations, were beginning to feel the need for improved fighter escort in their raids over Europe. General Arnold himself was subjected to public pressure as a result of heavy bomber losses. In 1942, Maj. Gen. Carl "Tooey" Spaatz, Eighth Air Force Commander, believed that the bomber forces' firepower and formation tactics negated a requirement for fighter escort under virtually all conditions. His successor, Ira Eaker, also was confident initially that the bombers could operate successfully beyond the range of fighter escort. In the fall of 1942, he had written to General Arnold that "three hundred heavy bombers can attack any target in Germany by daylight with less than four percent losses." But bomber losses rose sharply in December and January, prompting General Eaker to press Arnold for help in fixing the escort problem. Kepner's selection to command VIII Fighter Command was, at least in part, a response to Eaker's plea.

By April 1943, Kepner had a clear set of marching orders from the Commanding General of the Army Air Forces. As he recalled during an interview in the summer of 1944:

[the bombers] needed protection, needed it well organized, and a long way in, and *right now...*

ALL THE WAY TO BERLIN

> General Arnold sent me over here; said he wanted it set up and he was going to leave me to do it. How it was done was up to me, but there was no doubt in my mind that he wanted this escort.

Kepner's natural impatience, a trait hardly dimmed by his promotion to major general on April 27, drove him to see how much progress could be made on the escort problem even *before* he shipped overseas. He thought it possible that the P-38 could accompany bombers all the way to Berlin and that the P-51 had the same potential. On his own initiative he visited the companies that built the P-38, P-51, and P-47, and told the company presidents that they had to increase the range of their fighters. "I told them that I wanted to get to Berlin and had to have more gas in these planes." His insistence on range extension had results, but they were slow in coming.

In August 1943, Kepner arrived in England, proceeding to Watford and his new headquarters, a converted hotel called Bushey Hall. Here at the "front office" (known to the flying crews as "Ajax") he found the situation worse than he had expected. In the first place, the North African campaign (Operation Torch) had drawn off the bulk of American fighter resources in England. Kepner's predecessor, Brig. Gen. Frank O'D. "Monk" Hunter had consequently husbanded his resources. Despite increasing pressure, Hunter was reluctant to use his small force except to protect the bombers during their initial penetration and withdrawal. The few American pursuit craft available were P-47 Thunderbolts. Know as "Jugs" because of their unusually shaped fuselage, the eight-ton P-47s had a maximum combat radius of approximately 1754 miles. Still, General Hunter had not pushed for the longer range P-38s, preferring first to give the P-47 a complete trial in combat. To Kepner's feisty nature, this showed a decided lack of aggressive thinking.

At the time, VIII Fighter Command had only six groups of P-47s divided into two wings. Three of the groups were training while the remainder were combat operational. These units had been providing limited penetration and withdrawal support close to the French coast since their first escort mission in May 1943. Ninth Air Force fighters also were called upon to fly escort beginning in late 1943, and continued to support VIII Fighter Command until shortly before D-day in June 1944. Kepner acknowledged that short escort missions had some value: they provided combat experience for "green" pilots and demonstrated tactical ideas that would prove useful once long-range escort was begun. But short missions were not good enough, and no one knew this better than the bomber people themselves. In an appeal to VIII Fighter Command for more and deeper escort, Bomber Command's Maj. Gen. Fred Anderson had written, "It is obvious that the ideal fighter protection is that which can accompany the bombers from enemy territory to [the] target. Failing that, the greater the escorted penetration the better." To the new fighter boss, the urgency of this message and Fighter Command's failure so far to deal effectively with the problem were equally apparent.

Kepner picked his whole organization up like a rug and gave it a shake. He put combat-experienced people into staff positions; thus, there was never "the combat pilot's viewpoint" and "the headquarters viewpoint." They were one and the same. In the ever-shifting tactical environment of the European theater, this closeness of staff and line people turned out to be crucial in disseminating the "latest word" to all aircrews.

The Air Technical Section that Kepner established had been feverishly at work since January 1943 trying to provide a British source of external auxiliary fuel tanks for the P-47s. They tried a number of tanks that winter, but leaks, pressurization problems, and the British steel shortage frustrated

MAKERS OF THE UNITED STATES AIR FORCE

How the Guns Are Mounted on America's Big Planes

The Consolidated Liberator B-24D

The Boeing Flying Fortress B-17F

The reason why American four-motored heavy bombers have shot down German fighters by the score lies in the power and disposition of the defending guns. Until this week it was impossible to go beyond this bald and unsatisfactory statement in explaining to the public the outstanding successes of the United States Army Air Forces. The censorship forbade giving the position of all the guns or referring to the latest models by the complete designation figures.

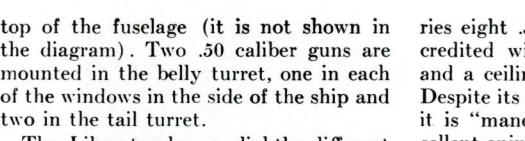

This week NEWSWEEK received the first copy to arrive in the United States of a booklet called Aircraft Identification, prepared by The Aeroplane, the authoritative British aviation magazine. This issue of Aircraft Identification is devoted to American planes and gives full information on the latest Boeing Flying Fortress, the B-17F, and the Consolidated Liberator, the B-24D. Since the booklet is on public sale in London, the censorship has been lifted.

The accompanying sketches from Aircraft Identification show front, bottom, and side views of the two American bombers with the gun positions indicated. The Fortress mounts a total of thirteen guns. In the nose it carries one .30-caliber and three .50-caliber machine guns. The turret behind the cockpit mounts two .50s and another is placed farther back on the top of the fuselage (it is not shown in the diagram). Two .50 caliber guns are mounted in the belly turret, one in each of the windows in the side of the ship and two in the tail turret.

The Liberator has a slightly different armament. The nose contains four .50-caliber weapons. Turrets behind the cockpit and in the tail each mount a pair of .50s, although in some cases British turrets with four .303 guns each have been installed. A belly turret with two .50-caliber guns completes the armament. This array of guns gives both the Fortress and Liberator heavier armament than any other known type of bomber.

Aircraft Identification gives the B-17F a speed of 290 miles an hour, a reduction of 30 miles an hour over earlier models with less armament. The Liberator is rated at 280 miles an hour. Both bombers are estimated to have only small bomb capacity—2½ tons for the Fortress and 3 tons for the Liberator. External racks may increase this tonnage but will cut down speed. However, too much larger versions of the two planes, the Consolidated B-32 and the Boeing B-29, are under construction.

Two new American fighter planes are singled out by Aircraft Identification for special praise. One is the 2,000-horsepower Republic Thunderbolt, which carries eight .50-caliber machine guns. It is credited with 390 miles an hour speed and a ceiling of better than 40,000 feet. Despite its great weight of 13,500 pounds, it is "maneuverable and has gained excellent opinions from RAF fighter pilots." Furthermore, it "gives promise of becoming one of the great airplanes of this war."

The other fighter is the North American Mustang. With a 1,150-horsepower Allison engine, the Mustang carries eight .50-caliber machine guns and with a speed of 370 miles an hour is "the fastest and best American fighter delivered to the RAF." For ground support and reconnaissance the Mustang "is excellent ... is very maneuverable and is sweet on the controls." It is not yet supercharged for high-altitude work.

Aircraft Identification also reveals the startling armament the British have installed in the Havoc, which is the two-motored Douglas attack bomber turned into a night fighter. In a long black nose on this machine the RAF has mounted a battery of twelve .303 machine guns. Another success for American planes has been chalked up by the Grumman Wildcat naval fighter (called the Martlet by the British). Aircraft Identification termed the Wildcat "the fastest airplane in service with the Royal Navy."

(Photo & Text: Newsweek, Oct. 26, 1942.)

ALL THE WAY TO BERLIN

Buy War Bonds - to Have and to Hold

MAKERS OF THE UNITED STATES AIR FORCE

Looking at flying range from l. to r.: Col. Stewart Towle, VIII Fighter Command Chief of Staff; Maj. Gen. William Kepner; Col. Francis Griswold, Ass't. Chief of Staff; and Col. Robert Burns, A-3. *(Photo: The Kepner Collection)*

their efforts. By summer, however, Air Technical Section began having some success, providing 75-gallon, 108-gallon (for which Kepner personally interceded with the British), and 150-gallon tanks through the remainder of 1943.

The new centerline tanks were thought by the pilots to look rather ungainly. They called these extra fuel tanks "babies." Skepticism of the flyers notwithstanding, the babies did the job. P-47 escort range increased dramatically. With no tanks, combat radius had been 175 miles; with a sing 75-gallon tank it jumped to 280 miles; with one 108-gallon tank it reached 325 miles, and with two external tanks 450 miles. Soon these tanks, General Kepner exulted, " . . .

[permitted Thunderbolt operations] within sight of Berlin, and that is a long step from the 175 miles, with the same damn airplane and the same boys flying it." The problems of pressurization, material shortages, and initial lack of support from U.S. suppliers had been overcome by persistence of Kepner and his staff.

Air Technical Section complemented the auxiliary tank program with other P-47 range and performance refinements. They replaced the Pratt and Whitney R-2800-21 engines with the -59 and -63 water-injected versions; they perfected fuel tank jettison systems, and installed a highly efficient paddle-bladed propeller.

While General Kepner applauded these

technical improvements, he was keeping the pressure on in other areas to lick the long-range escort problem. Fearful bomber losses suffered in the August and October 1943 Schweinfurt-Regensburg raids, due primarily to the lack of long-range fighter escort, had lent heightened urgency to the task. These combined raids had resulted in 95 of 441 aircraft lost on the August strike, and 82 of 257 on October's mission. Kepner urged his fighter groups to protect the bombers at all costs. A sign he posted in every briefing room read, "We have two scores we are aiming at—first the number of bombers we bring back safely, and second, the number of German fighters we destroy."

Under Kepner, air commanders were encouraged to experiment with new tactics. He insisted on "skull sessions" between pilots and planners. One early product of this procedure was refinement of the relay escort system that had been started under Monk Hunter. By late 1943, one group of fighters was escorting the bombers in their initial penetration of enemy territory, then a second group picked up the bomber stream and escorted it to the target where a third group would take over. Successive groups of fighters accompanied the bombers on their withdrawal. These tactics required thorough coordination and navigational skill, but they permitted the fighters to proceed straight to the rendezvous, thus extending range and minimizing the time spent weaving across-course above the much slower bombers. After October 1943, flight leaders also could get radar vectors to rendezvous points, courtesy of the new British "Y" Service.

Satisfied that substantial improvement in P-47 capability was under way early in the fall, Kepner turned his attention to bringing

ALL THE WAY TO BERLIN

the longer range P-38 Lightning and P-51 Mustang into the theater. At the end of September, he gave first priority to assembling P-38s that had arrived in England by surface transport. By October 15, some of these aircraft were already in action. Initially, the few P-38s available to accompany bombers all the way to the target were greatly outnumbered, but as their numbers increased during November and December, the P-38s were holding bomber losses in the target area to a supportable level for the first time since deep penetrations had been halted in the wake of the October Schweinfurt mission. With the arrival of Lightnings, General Kepner later recalled, "We began to think . . . that we were in the long-range escort business." Like the P-47s, the P-38s were outfitted with external tanks that improved their already superior range from an untanked combat radius of 260 miles to 520 miles with two 75-gallon wing tanks, and then to nearly 600 miles with two 150-gallon tanks.

The P-51 Mustang was slower arriving in the theater. It was early December 1943, before the P-51 became operational in Europe. General Kepner had insisted that Mustangs would be the only satisfactory answer to truly long-range escort. General Arnold had asked the RAF's Air Chief Marshal Portal to put his P-51s at the disposal of VIII Fighter Command for long-range escort. In January 1944, the RAF did provide four Mustang units. Meanwhile, Kepner was angry that P-51s coming from the States were slated for Ninth Air Force. The Eighth and Ninth Air Force Commanders reached a compromise: the 354th Fighter Group, first ETO American unit before 1944 to get the Mustang, was operationally assigned to VIII Fighter Command even though it belonged on paper to Ninth Air Force.

An element of two P-47Ds in WWII formation. *(USAF Photo)*

Originally built by North American Aviation for the British, the Mustang did its job extremely well. Incredibly, it had gone from design to first flight in less than 100 days. Pilots who flew the Rolls Royce Merlin-powered version agreed that the Mustang was the world's finest fighter plane. Its awesome basic fighting radius of nearly 500 miles could be augmented by external fuel to 850 miles. This once unheard of endurance for a pursuit aircraft was a quality referred to by crews as "seven-league boots," and it signaled the final push to develop an indomitable long-range fighter capability.

By any standard, this rapid rise in fortune for long-range escort during General Kepner's first four months of command was remarkable. In September the P-47s had been limited to 175 miles, and by November they ranged some 450 miles from England. The advent of P-38s in October and P-51s by December heralded a turnaround in the air war. The combined offensive punch of Eighth Air Force, Ninth Air Force, and the RAF fighter arms had indeed done the trick, changing the momentum of the entire war. At VIII Fighter Command, Kepner had driven his people—staff, engineers, flyers—to get the most out of the weapons available in the theater at any given time. According to General Kepner, "if it can be said that the P-38s struck the Luftwaffe in its vitals and the P-51s are giving it the coup de grace, it was the Thunderbolt that broke its back."

* * * * *

MAKERS OF THE UNITED STATES AIR FORCE

A lot had changed since the departure of most VIII Fighter Command aircraft to North Africa in early 1943, and since the peak bomber losses of that fall. Allied statistics echoed the transformation. As bomber claims of enemy aircraft destroyed dropped steadily after October, fighter claims mounted. Enemy fighters were not getting through the fighter screen to hit the bombers.

This shift had taken place during a period in which orders from General Eaker at Eighth Air Force were to "stick with the bombers" and not to wander off challenging the German fighters. But General Kepner had always wanted his fighters to seek out the Germans. Colonel Zemke's 56th Fighter Group had, in fact, been doing just that despite the official ban on leaving the bombers in order to forage for Germans. According to Zemke, his people pioneered a roving tactic in which ". . . the attempt was made to disrupt the enemy before he cold launch an attack. The immediate results were apparent on the score card—a sudden spurt of air victories for the 56th." The idea was to split the escort into a "seeking" unit that swept the skies ahead of penetrating bombers or behind withdrawing bombers while the rest of the escort provided continuous close flank cover. Finally, said Zemke, "the tactic leaked and I was invited to a dinner at General Doolittle's—with Kepner and a number of others in attendance." They pressed the young group commander for details, and a new aggressive escort policy was born. Doolittle, General Eaker's successor at Eighth Air Force, cleared the fighters to range away from the bomber formations and seek out the enemy.

Editor's note: AQ Volume 8 - Number 4, carried a full color picture on page 291 of a 413 series P-51D, without the dorsal fin. Mr. Anderson wrote AQ and it developed into this story within a story.
— GES —

"WILLIE Y" ANDERSON'S "SWEDE'S STEEDS".

In the early 1940s, "Willie Y" went through U. S. Army Air Corps flight training. Like so many of his peers he became proficient in P-40s, P-39s and the Lockheed P-38, before getting the opportunity to see combat in the P-51A, B and D derivatives. Lt. Anderson was assigned to the 353rd Fighter Squadron - "The Fighting Cobras" - in January 1943, in Tonopah, Nevada. The

Anderson chasing a V-1 Buzz Bomb (left edge of photo). *(Photo by: Willie "Y" Anderson)*

ALL THE WAY TO BERLIN

Doolittle's decision also reflected the offensive spirit of General Arnold's New Year (1944) message to his commanders: "Destroy the enemy Air Force wherever you find them, in the air, on the ground and in the factories." Kepner was quick to see that new opportunities to exploit fighter flexibility were on the horizon. On January 17, 1944, he issued a prophetic message to the pilots of his command:

> a fighter pilot must be able to use his versatile weapon in whatever way will do the greatest damage to the enemy . . . high or low, near or far, protecting bombers, destroying enemy fighters, preparing the way for our advancing ground troops, cutting the supply lines, strafing airdromes and other necessary missions. . . . Be ready. Today we are flying high altitude escort for heavy bombers. Tomorrow . . . ?

That offensive philosophy suited the fighter-pilot temperament. And, as the message had forecast, VIII Fighter Command's turn to the offensive did not stop with bomber escort.

Col. Glen Duncan, Commander of the 353d Fighter Group, started something new early in 1944. He led his flight in a strafing attack on a German airfield as they were returning from what had been an uneventful escort mission. One pilot described this armed buzzing as "roaring down at terrific speeds on a chosen object, zooming over it with inches to spare—and the closer the better—add the hazards of flak and ground fire and you have a sport that is practically irresistible." Duncan's experiment soon grew to be a major (and unauthorized) tactic in VIII Fighter Command.

In his own way, Kepner kept his superiors informed:

"WILLIE Y" ANDERSON'S "SWEDE'S STEEDS" *(continued).*

353rd had been activated on 15 November 1942, at Hamilton Field, California, under CO, Major Owen M. Seaman. While on the west coast, Willie recalls, General Kepner would occasionally sit-in on a few hands of poker with the young pilots. The 354th Fighter Group, (included the 353rd Squadron) was sent to England, where "Willie Y" was destined to be a fighter ace with 9 1/2 kills to his credit. Three air victories were scored in a triple kill on one flight, for which he was awarded the Silver Star.

Not counted amongst his victories are the German Buzz Bombs (V-1) he downed, being among the first Allied pilots to do so. "Most of the time, we stayed away from the things because every gunner on our side of the lines were always blasting away at them, and didn't much care if an aircraft was flying near them or not." Another one of "Willie Y's" victories — not on the scoreboard — was a Spitfire. "He shot at me first," Anderson noted, although feeling it was piloted by a german flyer.

During the war, Willie flew three different P-51s — all coded FT T and each bearing the name "Swede's Steed". Another pilot lost his first P-51B, but "Swede's Steed II" (Willie was a native of Sweden"), was the fighter he made his first air-to-air combat victory in. Inevitably, that aircraft was eventually relegated to training missions and a P-51D became "Swede's Steed III" for the remainder of his combat tour.

"Willie Y" became a commercial pilot after the war and covered the gamut from DC-3s to the Boeing 747. He retired from United Airlines in the 70s, but not from flying. AQ

Lt. "Willie Y." Anderson

- GES -

Spring 1989 • AVIATION QUARTERLY • **195**

MAKERS OF THE UNITED STATES AIR FORCE

> Once when my VIII Fighter Command was "tearing into anything" in Germany en route to or from the bomb target, I was riding in an auto with Spaatz. I said, "General, my Fighter Command is doing some screwy things en route and returning, but I think they pay dividends." Tooey Spaatz laid his hand on my knee and replied, "Yes Bill you are right, and so long as they pay dividends you will have a job." I got the point.

Kepner decided this "unorganized guerilla warfare" needed a dash of method. In late March 1944, he organized a unit that became known as "Bill's Buzz Boys." Volunteers from groups throughout the command gathered at Metfield for training by Colonel Duncan. The unit's mission was to develop airdrome strafing tactics that would be used throughout VIII Fighter Command.

Its mission completed, the unit was dissolved on April 12, 1944. Shortly thereafter, Kepner and his planners came up with the "Chattanooga Choo-Choo" and "Jackpot" strike missions against railroads and airfields respectively. On a single mission, as many as 1,000 fighters might be assigned. Each fighter group was given one of fifteen geographical sectors in Germany consistent with the range of aircraft assigned to the group. By the late spring of 1944, Ninth Air Force units also were doing armed reconnaissance on a large scale.

The effects on Germany of this concentrated strafing were devastating. In April alone, VIII Fighter Command strafers claimed 1,791 German aircraft destroyed and over 1,000 others damaged. Enemy ground movements in daylight were all but halted in many areas, and the enemy had to resort to special camouflage, dispersal, and

Above: P-47D on pierced-steel-planking (PSP). *(Photo: USAF)*
Below: An FW190A-8 with "defense bands" on fuselage during last days of war. *(Photo: USAF)*

ALL THE WAY TO BERLIN

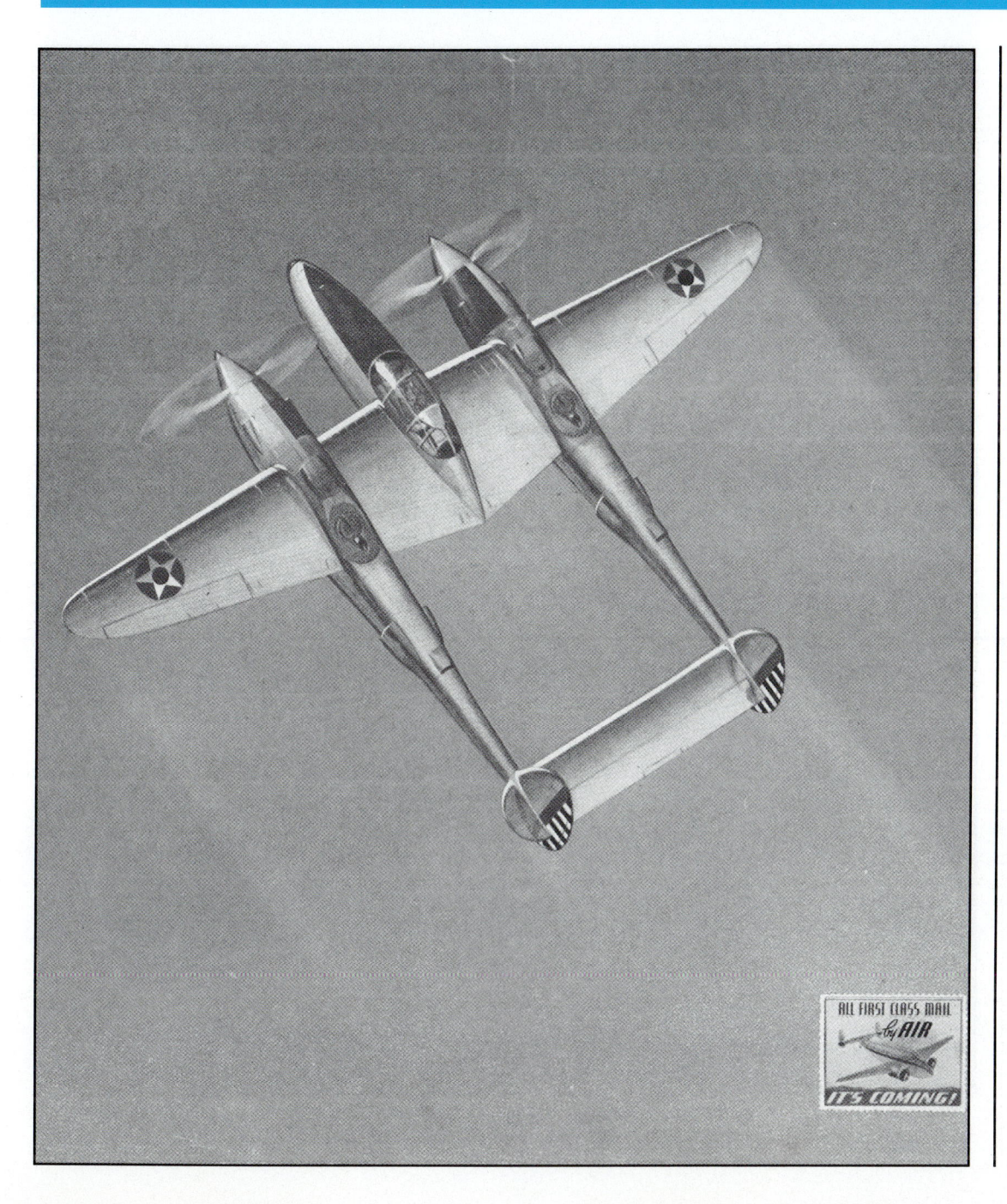

night techniques. Captured German documents attested to the effectiveness of strafing by Kepner's pilots and their Ninth Air Force colleagues, reporting that even motorcycles and isolated soldiers were attacked. After the war, Germany's fighter commander, Adolph Galland, said: "Nowhere were we safe from them; we had to skulk on our own bases. . . . [It was] a logical extension of tactics, permitting fighters to leave bomber formations to seek out the enemy."

Long-range escort, which made possible the first daylight raids on Berlin in March 1944, had been so effective against the Luftwaffe that in April Allied fighters were routinely leaving the bombers to destroy German planes on the ground. On May 20, three weeks before D-day, Kepner's fighters were cleared to fly strafing missions against enemy transportation in occupied France. Because General Kepner had had the foresight to organize and disseminate new tactics developed by his best pilots, the VIII Fighter Command was ready for a wide range of tasks to support the landings at Normandy in June 1944.

Kepner's fighters were even prepared to drop bombs on D-Day. According to Col. Hubert Zemke, Commander of the 56th Fighter Group, as early as November 1943, his P-47 pilots were dropping bombs in level flight behind formations of B-24s flying at medium altitude. They later practiced rudimentary dive skip bombing too. As in the strafing experiment, each unit sent a representative to learn these techniques and bring them back to his outfit. Fighter bombing was used to some extent by VIII Fighter Command, but it never became a major activity as it did for Ninth Air Force fighter bombers.

In each extension of its combat function,

MAKERS OF THE UNITED STATES AIR FORCE

VIII Fighter Command reflected its commander's belief in tactical flexibility as the key to fighter effectiveness. Each venture paid off because it was thought out in terms of the enemy threat and was tirelessly followed up to see if it was working. Kepner hounded logisticians and suppliers, but granted combat tacticians free reign to experiment and innovate. Ideas were not judged on who thought them up; if they were good, they were used until they no longer worked. It was this trait of tactical open-mindedness that helped make VIII Fighter Command an effective combat organization. Wing Commander Nigel Tangye of the RAF claimed some three weeks after the Normandy invasion that in his opinion one of the most remarkable achievements of the American air forces was the flexibility of Eighth Fighter Command. "No other command—RAF or USAAF," he said, "has ever been asked to mix the strategic with the tactical in such precise terms."

D-day itself had seen VIII Fighter Command fly a prodigious number and variety of sorties. Its P-38s provided a large measure of the cover given the nearly 6,500 vessels transporting the assault force. The P-47s and P-51s, together with Ninth Air Force fighters, provided a solid curtain of air superiority. Kepner put it this way:

> We formed a screen across the English Channel to the east of the surface vessel traffic that went clear down around 50 miles south of the beachhead, across the Cherbourg peninsula, and back across the Channel to the English coast—a half circle, and we maintained that thing from five minutes before first light on D-day until after 11 o'clock that day, solid, so that not a single damn German airplane came through there to go to the beach.

P-51 Ground Crewman helps to attach wing tank for farther air penetration of fighter coverage deep into Nazi held Europe.
(Photo: USAF)

198 • AVIATION QUARTERLY • Spring 1989

ALL THE WAY TO BERLIN

A P-51 being loaded with 500 lb bombs for another mission over fortress Europe.
(Photo: USAF)

In a ground strike role, too, General Kepner's men helped bottle up German road and rail activity within fifty miles of the front, sharing the interdiction task with the Ninth Air Force. Hermann Goering declared that "the Allies owe the success of the invasion to the air forces. They prepared the invasion; they made it possible; they carried it through."

For the airmen, the invasion itself was almost anticlimactic: "If you see fighter aircraft over you, they will be ours," General Eisenhower told his invasion forces. And that is the way it was. Some observers believe the air war was won in March and April 1944, during the peak escort and fighter ground-attack effort. Regardless of when the actual victory occurred, General Bill Kepner's versatile VIII Fighter Command played an important role in defeating the Luftwaffe. Following the invasion, Eighth Air Force bombers continued their deep thrusts into Germany, and VIII Fighter Command, with nearly half its strength now in P-51s, provided increasingly effective cover for the bombers

* * * * *

Early in August 1945, Kepner turned his office at Bushey Hall over to Brig. Gen. Francis "Butch" Griswold and took command of the Eighth's 2d Bombardment Division. The VIII Fighter Command in its brief ETO existence had flown more than 137,000 sorties and lost almost 1,300 pilots and planes, but it had destroyed or probably destroyed some 4,500 German aircraft and damaged 2,400 more. Bill Kepner could feel content about the job his once frustrated and struggling command had done. It had played a dominant role in what he himself knew was "... the greatest offensive fighter battle ever fought."

Hitler's Luftwaffe chief, Hermann Goering, said after the war that without long-range fighter escort, the Allied air offensive never would have succeeded. General Kepner remembered hearing Goering's opinion long before it became part of the official record of interrogations:

> When Germany surrendered at Rheims, France ... General Cannon, General Vandenberg, and I were waiting in the adjacent room for orders, and in case Germany refused our terms, we would start air operations immediately. When everything ended okay, General Spaatz came out laughing and said to me, "Bill, I asked Reich Marshal Goering, 'When did you know the jig was up?' Goering replied, 'When I saw the American fighter planes knocking down our German fighter planes over Berlin, I knew you could protect your American bombers and would bomb us out of the war.'" General Spaatz continued, "I understand you had some problems getting our fighter planes over Berlin, so I pass the compliment to you."

When Kepner assumed command of the 2d Bombardment Division in mid-August, he did not say farewell to VIII Fighter Command units. In September, Eighth Air Force Commander General Doolittle distributed Fighter Command's units among the Eighth's three bombardment divisions. Doolittle's objective was to more closely integrate fighter and bomber control, and to that end, Kepner's 2d Bombardment Division got five of the VIII Fighter Command's fifteen P-47 and P-51 groups. After the reorganization, General Kepner's new command consisted of some 900 B-24 bombers and more than 500 fighters. It operated in much the same fashion as a "numbered" air force and conducted bombing operations against Germany until V-E Day in May 1945. Under Kepner's leadership, the 2d Bombardment Division radically improved its visual bombing accuracy and experimented with a number of radar bombing techniques. The fighters continued to provide both escort for the bombers and offensive and harassing operations.

As the war approached its end, General Kepner was given command of Eighth Air Force, a job whose earlier distinguished incumbents were Generals Spaatz, Eaker, and Doolittle. Kepner's main concern following the surrender was management of Eighth Air Force withdrawal from the European theater. As that air force passed from the scene in August 1945, he took over the reins of Ninth Air Force, in whose care all duties of the Occupational Air Forces were left—aerial policing, photomapping, organizing air bases, and supporting the military government. By December, the process of demobilization was sufficiently advanced that all Ninth Air Force duties were transferred to the XII Tactical Air Command, with General Kepner as its new boss.

In January 1946, Bill Kepner finally returned to the United States. Nearly two and a half years had passed since General Arnold sent him to the European Theater. It had been a long war. But typical of Kepner's service history, he did not stay put for long. Within the month, he was named Deputy Commander for Army and Navy Aviation, Joint Task Force 1. Operating directly under the Joint Chiefs of Staff, this organization

B-24 Liberator:
For the pilot an unforgiving handsfull.
(AQ Collection)

ALL THE WAY TO BERLIN

P-51s on ramp. *(Photo courtesy of Robert J. Roskuski)*

was responsible for "Operation Crossroads," America's nuclear testing in the Pacific.

Nine months later, Kepner traded the middle of the ocean for the middle of the American continent, arriving at Scott Field, Illinois, to head up Air Technical Training Command. There he remained until October 1947 when he was assigned to the office of the Deputy Chief of Staff for Research and Development at Air Force Headquarters. Within the deputate, Kepner held several positions, all directly responsible for Air Force nuclear weapons and programs. In August 1948, he left Washington to take command of the Air Proving Ground at Eglin Air Force Base, Florida.

There was one last operational command in store for Bill Kepner. He was promoted to lieutenant general in June 1950, and named Commander in Chief of the unified Alaskan Command, succeeding his old Air Corps Tactical School classmate, Nathan Twining. Kepner's early years with the Army and Marine Corps made him remarkably well fitted to lead a tri-service command. Since 1947, when the Joint Chiefs had put Alaska's defense in the hands of a senior Air Force officer, F-80, F-82, and F-94 jet interceptors had stood watch against possible Russian air attack. Kepner's Army and Navy units in Alaska had similar defensive duties for their own special provinces of land and sea. Conditions in Alaska's muddy summers and devastating winters were primitive, but "if the enemy invades," said Kepner, "we'll hand him quite a jolt."

In December 1952, about three weeks before his sixtieth birthday, General Kepner relinquished his post to Maj. Gen. Joseph Atkinson and returned to Washington for a retirement ceremony on February 28, 1953, at Bolling Air Force Base. His more than forty years of military service spanned the period from America's tentative first steps in military aviation to the creation of nuclear weapons. He died at Orlando, Florida, in 1982 at the age of eighty-nine.

General Kepner's career as a member of three military services—and his work as a lighter-than-air pioneer, stratosphere explorer, and fighter commander—is unique in the annals of Air Force history. His technical and tactical skills and his willingness to experiment in both areas earned him a place among the leaders in the development of air warfare.

Spring 1989 • AVIATION QUARTERLY • 201

MAKERS OF THE UNITED STATES AIR FORCE

Author

LIEUTENANT COLONEL PAUL F. HENRY is Commander of the 335th Tactical Fighter Squadron, Seymour Johnson Air Force Base, North Carolina. The 335th "Chiefs" trace their history from RAF 121 Squadron, one of the three original American Eagle Squadrons of World War II. In 1967, he earned a Regular commission and a bachelor's degree in humanities from the U.S. Air Force Academy, where he won the Hall Nordhoff Trophy for achievement in English. Following pilot training, he flew 162 combat missions in Vietnam with the 315th Special Operations Wing. He earned a master's degree in English at the University of Connecticut and an M.Ed. from Auburn University, and has been a member of the Air Force Academy Department of English. His articles have appeared in several military and educational journals.

Sources

The core source for this biographical sketch is the Kepner Document Collection, housed at the USAF Historical Research Center, Maxwell AFB, Alabama. The collection contains hundreds of personal and official documents along with a number of tapes, manuscripts, photographs, and other memorabilia given to the Center by General Kepner. These are invaluable in gaining a sense of the man and of the impact he made on his surroundings. Several telephone interviews with the General and a series of letters to the author written between January 1981 and July 1982 were instrumental in clarifying material in the collection and in gathering personal anecdotes.

A few general works provide background information pertinent to the various periods of Kepner's career. Most useful among these is the *Army Air Forces in World War II*, edited by W. F. Craven and J. L. Cates (University of Chicago Press, 1951; Reprint, Office of Air Force History, 1983). Volume II, *Torch to Pointblank* (1949) and Volume III, *Europe, Argument to V-E Day* (1951) are especially valuable. Other general sources include DeWitt S. Copp's *A Few Great Captains* (Doubleday, 1980); and *The Mighty Eighth* by Roger A. Freeman. Copp's *Forged in Fire* (Doubleday, 1982) examines WW II Air Force operations with considerable detail on the Combined Bomber Offensive and the long-range escort problem.

By far the best source of "atmosphere" from the fighter operations standpoint is Grover C. Hall's *1000 Destroyed: The Life and Times of the 4th Fighter Group* (Aero Publishers, 1978). In several letters to the author, Colonel Hubert "Hub" Zemke provided first-person operational accounts that also help capture the flavor of VIII Fighter Command.

Several 1930s' issues of *Scientific Monthly* and *The National Geographic* aid in reconstructing the "Explorer I" stratospheric expedition, but the major source on this topic is an oral history interview from the Kepner Document Collection. *Air Force Magazine* (September 1978) and *Aerospace Historian* (September 1971) contain lengthy articles that survey Kepner's balloon career. His manuscript entitled "Riding the Storm in a

Maj. Gen. Kepner aboard the flagship *McKinley* during the 1946 Bikini atoll atomic bomb tests.
(Photo: The Kepner Collection)

ALL THE WAY TO BERLIN

National Balloon Race" also contributes details on ballooning experiences. *Life*, *Time*, and *Army, Navy and Air Force Journal* yielded a number of human interest items from General Kepner's service in VIII Fighter Command, the nuclear test program, and Alaskan Air Command.

Among many pertinent official documents several merit mention. The first of these, a very thorough unit history indeed, is *Achtung Indianer!: The History of the United States VIII Fighter Command* (East Anglia, U.K., 1944) by Lt. Col. Waldo Heinrichs. Other significant sources include *The Long Reach: Deep Fighter Escort Tactics* (VIII Fighter Command, 1944), and *Eighth Air Force Tactical Development: August 42-May 45* (Army Air Forces Evaluation Board, 1945).

AVIATION QUARTERLY

B-24 Liberator is built by Consolidated. Heavily gunned nose and tail turrets have been added to protect it during daylignt bombing. It is flown in all theaters of war and Navy uses it as a long-range patrol bomber.

AVIATION QUARTERLY

P-70 — Midnight Mauler is Douglas' splendid A-20 attack bomber rebuilt as a night fighter. Plastic nose has been closed to hold interception gear. Bomb bay is sealed and carries extra gas tanks. It mounts four cannons.

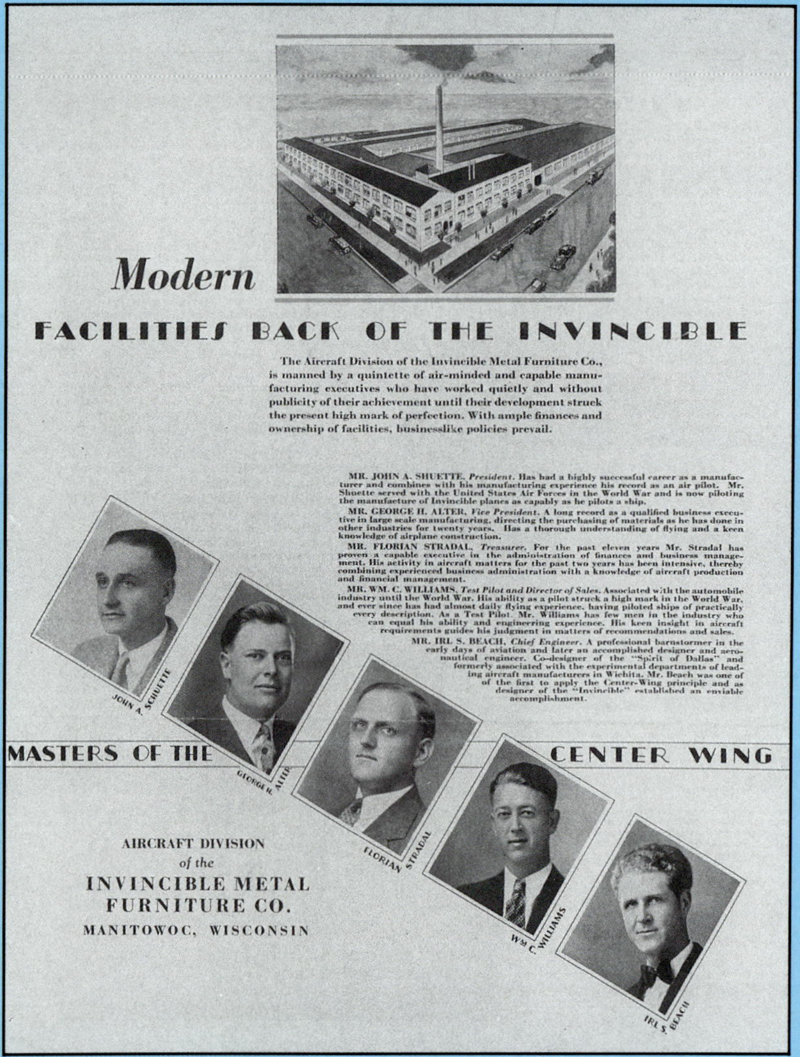

Modern
FACILITIES BACK OF THE INVINCIBLE

The Aircraft Division of the Invincible Metal Furniture Co., is manned by a quintette of air-minded and capable manufacturing executives who have worked quietly and without publicity of their achievement until their development struck the present high mark of perfection. With ample finances and ownership of facilities, businesslike policies prevail.

MR. JOHN A. SHUETTE, *President*. Has had a highly successful career as a manufacturer and combines with his manufacturing experience his record as an air pilot. Mr. Shuette served with the United States Air Forces in the World War and is now piloting the manufacture of Invincible planes as capably as he pilots a ship.

MR. GEORGE H. ALTER, *Vice President*. A long record as a qualified business executive in large scale manufacturing, directing the purchasing of materials as he has done in other industries for twenty years. Has a thorough understanding of flying and a keen knowledge of airplane construction.

MR. FLORIAN STRADAL, *Treasurer*. For the past eleven years Mr. Stradal has proven a capable executive in the administration of finances and business management. His activity in aircraft matters for the past two years has been intensive, thereby combining experienced business administration with a knowledge of aircraft production and financial management.

MR. WM. C. WILLIAMS, *Test Pilot and Director of Sales*. Associated with the automobile industry until the World War. His ability as a pilot struck a high mark in the World War, and ever since has had almost daily flying experience, having piloted ships of practically every description. As a Test Pilot, Mr. Williams has few men in the industry who can equal his ability and engineering experience. His keen insight in aircraft requirements guides his judgment in matters of recommendations and sales.

MR. IRL S. BEACH, *Chief Engineer*. A professional barnstormer in the early days of aviation and later an accomplished designer and aeronautical engineer. Co-designer of the "Spirit of Dallas" and formerly associated with the experimental departments of leading aircraft manufacturers in Wichita. Mr. Beach was one of of the first to apply the Center-Wing principle and as designer of the "Invincible" established an enviable accomplishment.

MASTERS OF THE CENTER WING

AIRCRAFT DIVISION
of the
INVINCIBLE METAL
FURNITURE CO.
MANITOWOC, WISCONSIN

AVIATION QUARTERLY

BASIC ENGINEERING PRINCIPLE
Revolutionizes
SPEED~POWER~STABILITY

Engineers who refused to take custom for granted have now developed one of the greatest contributions to the aircraft industry: "the perfected center-wing."

So logical, so unmistakably clear to any analytical mind is the principle of the straight line, that Invincible engineers never departed from that basic fact. Backed by exceptional aeronautical skill, resources and equipment, these engineers labored silently until accomplishment, far beyond their own expectations, drew the spotlight of admiration upon their achievement. A beautiful ship, with speed, power and superb performance!

WINGS are placed in alignment with the center of the propeller thrust, giving perfect balance and greater speed under all flying conditions. As definitely as a straight line is the shortest distance between two points, so surely does air leave a straight line faster. As positively as a straight arrow travels faster than one that is bent, does the "Invincible" out-speed any other wing design of equal power. Its racy lines, its dashing streamline effect is carried out in its very performance.

Wings of the semi-cantilever type are set at 1½ degree angle of incidence and 1 degree of dihedral which is adjusted at the upper end of the "V" struts, giving a radius of from 0 to 2½ degrees. Maximum lifting power thus accomplished insures quicker take-off.

The Invincible is powered by a 170 H.P. Curtiss Challenger motor, takes off in eleven seconds and develops the first thousand feet climb in fifty seconds; the second thousand feet altitude is attained in one minute with reserve power. A cruising range of 700 miles and a dependable cruising speed of 120 M.P.H. gives remarkably efficient travel for business flying. Top speed of 142 M.P.H. is attainable without overtaxing the power plant.

VISIBILITY

Sliding windows of non-shatterable glass give a wide range of visibility along both sides. Overhead window gives pilot clear visibility to the rear before taking off. This window is large, allowing sufficient space for auxiliary exit. Position of the front seats gives the pilot clear view forward in landing. Windows on each side in the bottom of the cockpit give the pilot a clear downward view.

CONSTRUCTION

The fuselage and tail are constructed entirely of chrome molybdenum steel tubing, all reinforced beyond standard strength requirements. Covered with sturdy fabric, treated with special pigment and finished with remarkable smoothness and high polish. The ship is 25'7" long and 7'3" high; wheel tread is 6'.

Wings are constructed of spruce with double system of internal bracing, employing the N.A.C. M 15 airfoil. All points of compression in the wings are of double steel tubing, except the root end which is a solid wood web, special fittings being employed to attach the tie rod bracing. The leading edge of the wing is covered with Duralumin. The covering is of fabric treated with special pigment. Wing area (including aileron) 228 sq. ft. The wing span is 40 ft.

Ailerons are of the unbalanced type. The leading edge is rounded to dovetail the recess in the wing, insuring perfect streamline in any position. Flight tests have proven this design to have exceptional lateral control. Aileron area is 22.6 sq. ft. Other areas: Rudder area, 8.1 sq. ft.; Fin area, 5 sq. ft., Stabilizer area: 12.2 sq. ft.; Elevator area, 13.4 sq. ft;

Luxurious COACH WORK
Unsurpassed for Passenger Flying

Roomy seats are anchored to the fuselage frame and beautifully upholstered to match the cabin interior. Low center of gravity accomplishes an attractive feature of entering the cabin as conveniently as a motor car. The large oval door facilitates entering and leaving the plane, and the accommodations give passengers luxurious flying comforts that encourage air travel.

The position of the door is back of the rear seats, giving a passage three and one-half feet back of the aisle. Luggage space does not interfere with entry or exit. The metal window frames and instrument panel are beautifully finished in grained walnut. Every detail is completed in thoroughness and style.

Price $7800.00 Flyaway Factory Airport

INSTRUMENTS AND CONTROLS

Tackometer, oil pressure gauge, Fahrenheit gauge, altimeter, carburetor choke, altitude adjustment, air speed indicator, switches for indirect lighting to instrument panel, navigation lights, compass and ignition switch are centralized on the panel.

Dual wheel type control side by side, either control quickly detachable. Hinge-type throttle for either right or left seat control. A stop is employed in the throttle eliminating the hazards of leaving the ship unattended while the motor is running. The throttle cannot creep open. Also prevents accidental opening of the throttle by passengers. Cables extend from a sprocket and chain hook-up to the ailerons. Cables are also used on the rudder and elevator controls. Navigation, power and lighting instruments all latest standards are controlled from the dash.

Two gasoline tanks, each of 30-gallon capacity, are located in the wings and are made of welded aluminum. These are insulated from their steel retainer strap with felt. Oil is fed to the engine by gravity from a 5-gallon tank located in the motor mount. The engine mount of the ring type is a special feature detaching the entire nose of the cabin, thereby facilitating the use of various motors. While the Curtiss Challenger 170 H.P. is standard equipment, other motors may be installed if desired.

The under-carriage is extremely sturdy and equipped with efficient shock absorbers, brakes and landing gear. Tail wheel facilitates efficient taxiing and ground handling. At a landing speed of 45 M.P.H. the under-carriage responds with remarkable smoothness and comfort so important to business flying.

INVINCIBLE FOUR PLACE *Center-Wing* **CABIN MONOPLANE**

(Compliments: History of Aviation Collection, Dallas, Tx.)

Dick Merrill, EAL captain.

DICK MERRILL, CLASSIC PILOT

by
Pat Pateman

Dick Merrill's desire to fly surfaced as he watched the famous Katherine Stinson perform acrobatics at a state fair in Mississippi in 1914. He was twenty at the time and working as a minor league baseball player. He served a short tour overseas with the Air Service in World War I but was disappointed when he did not receive flight training as promised by his recruiter. Shortly after his return he plunked down his life savings to buy a war-surplus JN-4. While on his way to Columbus, Georgia to pick up the Jenny, Dick dropped by San Antonio, Texas and learned to fly with the famous Stinson family. When he received his license, he joined the Gates Flying Circus and spent several years as a circus performer traveling with Gates around the country.

Merrill became an airmail pilot in 1927, flying the New York to Miami route in a Pitcairn Mailwing. He later became associated with Eastern Air Lines (EAL) and Eddie Rickenbacker, President and General Manager. It was this connection which led to his attempt to fly in a single engine airline airplane to Europe and back with just one refueling stop in London. He persuaded Harry Richmond, a well-known band leader of the era who had just purchased a surplus Vultee VI-A to make the crossing with him. The flight in the Vultee, named "Lady Peace" was accomplished in 1936 and although the two-way crossing of

Pitcairn PA5 Mailwing. *(Photo: Courtesy S. Pitcairn.)*

the Atlantic was completed successfully, the flight was not without incident. On the return leg, the duo ran into foul weather and Dick decided to drop below the overcast. He advised Richmond of his intentions by shouting, "We're going down." Harry obviously misread the statement and, having been briefed on his duties in an emergency, pushed the appropriate levers and began dumping the remaining fuel overboard. Fortunately, Dick was able to stop Harry before their entire fuel supply was depleted. Just enough fuel remained for Merrill to make landfall and set the plane down in a muddy bog, a few miles short from Harbor Grace airport in Newfoundland.

The flight, already newsworthy, had received additional press coverage at the outset when, at the suggestion of Sid Shannon and Eddie Rickenbacker, Dick and Harry loaded 10,000 ping pong balls into the wings and fuselage to help the plane stay afloat in the event of a forced landing on water. Although the "ping pong ball" experiment fortunately was never put to the test, Harry Richmond continued to reap publicity from the gimmick by distributing autographed ping pong balls after the flight to his many fans.

Dick Merrill continued flying for EAL, becoming the favorite pilot on the New York to Miami run as passengers fought to get on board his flights. His popularity stayed with him into 1952 when he became General Eisenhower's pilot during Ike's presidential campaign. Earlier, in 1947, as a promotion for long distance air travel, Merrill and Rickenbacker flew Arthur Godfrey and a plane load of dignitaries to South America in a new Constellation. Merrill was forced to retire from EAL in 1954 as he turned sixty.

He was still active at the age of 78 however when, with Captain Paul Slayden, he set a new speed record delivering a widebodied Lockheed 1011 Tri-Star jet transport from Palmdale, CA to Miami, FL. The flight was made at 39,000 feet in three hours and 33 minutes and a ground speed of 710 mph.

In 1970 Sidney L. Shannon began constructing an air museum at his commercial airport in Fredericksburg, Virginia. He wished to honor his father, the former Operations Manager of EAL by displaying his memorabilia to the public. Because of the friendship between his father and Dick Merrill, he invited Dick to be the curator. When the museum opened officially on Father's Day in 1976 it was an instant success. Crammed with airworthy, antique airplanes and memorabilia from Shannon and Dick Merrill it drew crowds continually.

Dick Merrill served as curator of the Shannon Air Museum until his death in 1982. **AQ**

This static-test version of the TriStar, girded with 300 load frames which, when attached to hydraulic jacks, received the equivalent of 15 years' hard service in less than two years of torture testing — one of the reasons that development costs are so high for such aircraft. (Lockheed California Company) *(Courtesy of Tab Books, Inc.)*

FREDERICKSBURG to RICHMOND

A MUSEUM IN THE MAKING

*by
Pat Pateman*

Aviation museums usually begin as a seed in some astute aviator's mind, waiting for the right time and the right place to sprout its wings. In the case of the Virginia Aviation Museum (formerly the Shannon Air Museum) the catalyst was Sidney Shannon, Junior a successful Virginia businessman. An ardent aviation buff in his own right, Sidney was searching for a permanent home for his father's aviation collection. His father, Sidney Shannon, Senior a chief operations manager for Eastern Air Lines (EAL) had retired to Fredericksburg, Va., bringing with him hundreds of historical gems he had accumulated during his years with EAL.

In 1970, young Shannon decided to build a facility on Shannon Airport in Fredericksburg, large enough to hold his father's memorabilia and his own modest collection of flyable, antique aircraft. Sid scheduled an air show on father's day in 1976 as a salute to his father's contribution to aviation. Thousands attended the air show to watch Bob Hoover, the Golden Knights and others and stayed to view the historic exhibits in the new facility. A museum was born.

Sid selected as curator for the museum, the colorful Dick Merrill, retired EAL pilot

Curtiss OX-5 engine display with propeller. *(Photo by: Marty Stanton)*

Gat-1 Singer IFR trainer. 1933 Aeronca engine display, VAM. *(Photo by: Marty Stanton)*

and a long time friend of his father. Dick arrived from Miami with his own extensive collection of aviation memorabilia. By the time the museum opened its doors to the public in 1976 it boasted one of the most complete repositories of commercial airline historical artifacts at the time. Besides the visual displays of uniforms, log books, and flight charts of old, the museum boasted hundreds of photographs and films of famous air pioneers and media celebrities who were pilots, such as Harry Richmond and Arthur Godfrey.

Sid Shannon's continued interest in classic airplanes led him to purchase original or replicas of airplanes Dick Merrill had flown to add to the display. The most unique purchase was a Vultee VI-A, similar to the one Dick Merrill had flown round trip to England with Harry Richmond in 1936. Since the original one, the "Lady Peace" had ended in a trash heap in Madrid, Sid settled for the only other Vultee available, which had been restored to mint condition by Harold Johnson. When the plane named the "Spirit of Pueblo," arrived at Shannon it was renamed the "Lady Peace" in honor of Merrill.

Sid died suddenly in 1981 following heart surgery in Texas. Although he had willed his memorabilia and classic aircraft to the newly formed Virginia Aeronautical Historical Society (VAHS), his will did not provide for a building to house the collection. The VAHS continued to operate the museum at Shannon Airport while the Sovran Bank, trustees for the Shannon family resolved the problems of Sidney Shannon's inheritance. By 1986, the Bank Trustees agreed to turn over the collection to the

VAHS as long as the Society could provide a secure facility for the collection within a short time frame.

The Society scrambled to satisfy the bank's requirement under the critical deadline. Thanks to the board of the VAHS and the innovative Neil November, prime acreage, located at the entrance to Richmond International Airport (formerly Byrd Field) Va., was made available to the VAHS for a museum. An extensive fund drive was immediately initiated. Contributions topped $500,000 and several local businesses donated the engineering and labor support to begin construction within months. The initial building named the Freelander, in honor of a major contributor was completed on schedule and the move from Fredericksburg to Richmond was made in the spring of 1987. The bulk of the collection of historical airplanes, many of which had lost their air-worthiness status during the legal hiatus were trucked to the new museum. The few remaining air-worthy classics were flown into Richmond International Airport by VAHS members in aviator's garb reminiscent of the particular aircraft's heyday. Other classics have found their way to the new Virginia Aviation Museum, to the extent that a year after its opening adequate space is at a premium. Further fund raising to support additional construction continues to be the highest priority of the museum board of directors.

Today the museum is operated by a minimum of paid employees and augmented by some 75 - 80 volunteers who attend the front desk, man the gift shop, operate the film theater and conduct tours of the museum on a daily basis. **AQ**

1932 Aeronca C-2 N Scout. Referred to as the "Flying Bathtub". *(Photo by: Marty Stanton)*

AVIATION QUARTERLY

This 1916 SPAD VII, one of only eight of this type remaining hangs from the ceiling directly above the front desk. The SPAD is considered one of the great fighter planes of all time. *(Photo by: Marty Stanton)*

AVIATION QUARTERLY

VAM

VIRGINIA AVIATION MUSEUM

by
Pat Pateman

Two boys burst through the glass doors and gaze up at a 1916 SPAD VII hanging from the ceiling at the entrance to the new Virginia Aviation Museum located at the entrance to the Richmond International Airport in Virginia. One of the volunteers drops what he is doing at the Gift counter and hurries over to "man" the admittance desk.

"You boys come here to see our airplanes?" Shirley Carter, the "desk" man can't resist smiling at them. Their upturned faces swing from side to side as they glimpse uncommon airplanes from an earlier age seemingly floating across the ceiling of the museum. The younger of the boys raises his hand up towards the World War I training plane hanging from the ceiling as if, with a big stretch he might touch the tire hanging down under the fuselage.

The father of the boys pays a nominal fee and receives a blue brochure in return, revealing the vital facts about the sixteen antique and classic airplanes, housed in what is named the Freelander wing of the Virginia Aviation Museum. Within a few seconds, the boys become deeply involved in learning the details of men and machines from an earlier age.

The above mural, painted on the exterior east side of the museum draws the aviation buff to the Freelander building with increasing numbers. *(Photo by: David Cooper)*

AVIATION QUARTERLY

That's what the new Virginia Aviation Museum (VAM) is all about: breathing new life into Virginia's impressive aviation history. The VAM, located at the entrance to the International Airport in Richmond, Virginia is owned and operated by the Virginia Aeronautical Historical Society (VAHS.) It houses an extensive collection of vintage and historically important aircraft. Included in the collection are the only known airworthy French SPAD VII fighter from World War One, the only surviving Vultee VI-A transport from the 1930's, a Pitcairn Mailwing, a Bellanca Skyrocket, and a Standard E-1 training plane from World War One. Also included in the collection are a 1939 Stinson SR-10G Reliant, a 1927 Travel Air 2000, a 1937 Fairchild F-24G, a 1953 Piper PA-18 Super Cub, a 1932 Aeronca C-2-N, and a 1929 Curtiss Robin.

One observer, Clayton O. Bailey on a visit from Buffalo, New York in March 1988 was very impressed. He said upon leaving, "I've visited the National Air and Space Museum, Silver Hill Garber Facility, the Air Force Museum in Dayton, the Canadian Warplane museum and others too many to mention. To me, this museum rates a 10 in my book. It's tops."

Besides the original Shannon collection which transferred to the museum from

*Photo above clockwise:
1914 Standard E-1 WWI trainer;
1928 Heath "Super Parasol;"
and Travel Air 2000.*

*Below:
1927 Pitcairn PA-5 Mailwing.
(Photo: Marty Stanton).*

216 · AVIATION QUARTERLY · Spring 1989

AVIATION QUARTERLY

The 1941 Bucker BU 133-C was considered the fines aerobatic aircraft for training and competition ever built up until World War II. This particular aircraft was once owned by Count Jose L. Aresti, originator of the Aresti Aerobatic Shorthand system. Painted to represent the color scheme used by the late Beverly "Bevo" Howard, one of the world's premier airshow aerobatic pilots who made the Jungmeister a well-known and sought-after aircraft. Howard's original Jungmeister is part of the display at the National Air and Space museum in Washington, D.C. *(Photo by: Marty Stanton)*

AVIATION QUARTERLY

Fredericksburg, Virginia in 1987 (See companion article, "A Museum in the Making"), the museum has attracted gifts of four other historic aircraft with more gifts being considered. When additional construction becomes a reality and permanent visitors galleries and administrative offices are built, the original Freelander building will take on its rightful task of restoring aircraft donated to the museum.

The museum belongs to the Society of course but has the State of Virginia's full backing. Ken Rowe, Director of the State Department of Aviation and a member of the Society's Board of Directors said both the society and the museum will build on the idea of commemorating the state's aviation heritage. To help in fulfilling that promise, Mr. Rowe has assigned two aviation educators (commercial flight instructors) to the museum on a temporary basis. The two educators are physically located in the museum and are developing programs to enhance the study of aviation within the school system. Tours of the museum by school children and such diverse groups as members of the Richmond Symphony Orchestra have now become commonplace.

The museum is also home to the Virginia Aviation Hall of Fame which boasts as its first inductee, Admiral Byrd, a Virginian. Since then the VAHS has added to

Above left:
The Vultee VI-A one of the most advanced single engine monoplanes used by the airline industry in the 1930's. *(Photo by Marty Stanton)*

Below left:
View of the 1,000 horsepower Wright Cyclone R-1850 radial engine installed in the Vultee owned by Wm. Randolph Hearst. *(Photo by Marty Stanton)*

A 1938 Stinson SR-10G Reliant. An example of combining utility and beauty, this strikingly handsome aircraft — with its distinctive "gull wing" — was known for its ease in handling and its ability to carry 4 to 5 people in comfort. The VAM Reliant is powered by a 300 hp Lycoming radial engine and is decorated in an American Airlines early paint scheme. *(Photo: David Cooper)*

1928 Bellanca CH-400 Skyrocket. *(Photo: Marty Stanton)*

A 1937 Fairdhild 24G. The VAM display is a deluxe model, with plush upholstery, rolldown windows, wing flaps, extra instruments, electric fuel gages, and a hand-rubbed finish. Restoration of the Fairchild was done in Virginia Beach, VA by Al Jenkins and flown to the Museum on 22 May 1987. *(Photo by: Dave Cooper)*

the list other Virginians who have met the criteria and deserve a place on the honored wall near the Admiral.

Other exhibits on display throughout the gallery consist of early century aero engines, navigation devices, flight clothes and precisely constructed model airplanes. A 75 seat theater, donated by J.D. Benn, a Virginia aviation pioneer is available to show the museum's extensive collection of historic and current aviation films and cassettes. There is also a viewing gallery atop the theater which overlooks the extensive airplane collection, half of which appear "airborne" from that vantage point. That is where camera buffs may photograph the airplanes as if they were actually in flight.

The museum boasts an extensive library, where personnel are now engaged in indexing the collection transferred from the Shannon Museum plus the hundreds of additional books donated by members of the VAHS. When this task is completed, the library will be available for serious researchers by appointment.

Now that its a going concern the museum publishes a calendar of events quarterly, advertising upcoming programs of featured speakers and film showings on a regular basis.

At present, the museum is open from 10 to 4 p.m. every day but Monday, and the predominately male volunteers are eager to guide you around the place, give you briefings on the different airplanes, their historical importance and when pressed, will chat with you about their own happy times in the cockpit.

Of course, it's not the largest Aviation Museum. It doesn't have space vehicles

AVIATION QUARTERLY

nor do yesterday's commercial aircraft noses protrude into the aisles. It does have however, practically "hands on" views of the cockpits of some very old and very well preserved classic aircraft. One stands close enough to smell the plush upholstery in the Fairchild 24 and can actually read the gas gauge in the Super Cub. The mail bag hanging out of the cockpit of the Pitcairn takes you back to the twenties when air devils in jodhpurs and leather helmets roamed the darkened skies providing the first ever airmail service.

Step outside the museum and you hear the familiar roar of approaching commercial airlines close enough to read their logos and catch glimpses of general aviation aircraft jockeying for space along the path to Richmond airport runway 20 to the south.

Visitors to this museum have reached VAM projections and growing larger as word of its collection of classic aircraft gets around. The good news is that you won't have to mortgage the car to park at VAM. The parking is free. Come by and see for yourself. **AQ**

Above left:
Proposed future expansion, VAM.
(Photo by: Marty Stanton)

Below left:
Virginia Aviation Hall of Fame
VAM, Richmond
(Photo by: Marty Stanton)

Ed Massey,
Vice President of VAM

Bill Kennedy,
Chief of Maintenance

Florence Haskings,
Library volunteer

Volunteers Shirley Carter and Art Wiggins greet the visitors at the entrance to the museum and conduct tours of the classic aircraft collection on request.

(Photos: Marty Stanton)

Melissa Seiler,
Educational Specialist

Betty Harris,
Asst Ed. Spec

Lt. Col. Yvonne C. Pateman, USAF (Ret.)

About the Author

Col. Pateman has written numerous stories for Aviation Quarterly. She is considered an expert on the history of women in aviation.

While still current as a private pilot, 'Pat' tours the speaking circuit as well.

Recently, she addressed an audience at the Air Force Academy, commemorating her donation of personal books, papers and aviation artifacts.

"What a thrill. There were also ten generals in the audience," she exclaimed.

The New-Day Barling NB-3 Sets New Standards

in Design — Performance — Economy — Structural Strength — Efficiency — Method of Manufacture

Designed—

— To meet the growing demand for a three-place, light plane, with new production dual ignition, air-cooled, 60 H. P. engine for quick take-off, high rate of climb, high top speed, low landing speed and a useful load of over six hundred pounds.

— To be produced by latest machine methods, for uniform, straight-line and quantity production, making standardized airplanes, for cheaper maintenance and lower operating cost, to eliminate as much as possible the hazards of human element.

— By an organization well known and established, amply financed, with 52,600 square feet factory space, equipped with modern machinery, modern jigs and dies, modern production methods, skilled labor, with engineering and designing personnel who are capable of keeping far in advance of present-day construction.

Results—

— A three-place monoplane with a newly-principled construction, based on new and proved fundamentals that carry the standard of aircraft far above the present-day basis of comparison. The New-Day Barling brings you a metal structure far stronger than any existing type used in aircraft, and yet with the enormous strength of this structure the weight is materially lessened. With the New-Day Barling highly-efficient wing, you have less weight, greater load-carrying capacity, higher rate of climb, greater stability, better vision and a higher speed than any three-place ship of its equal power available today.

— An ideal training ship, due to its sturdy construction, visibility, and its exceptionally low operating cost.

— A light airplane that does not require an expert pilot, as is the case with a majority of the light planes.

AVIATION QUARTERLY

The home of the New-Day Barling, comprising 52,600 sq. ft. of floor space.

THE PRESENT factory is equipped with 52,600 square feet of floor space in which the latest and most modern type of machinery has been placed for the stamping out of ribs, beams, and numerous other parts in addition to special jigs for each and every part of the ship from tail skid to engine mount. An overhead gas and oxygen system has been installed in order to eliminate time and loss in moving welding tanks. Straight-line production methods are employed throughout. One visit to the New-Day Barling NB-3 Plant will convince you of the efficiency and precision employed in the construction of this wonderful New-Day Plane.

We have produced for your inspection and approval a three-place ship weighing 690 lbs., as against the average OX-5 three-place ship weighing 1250 lbs., giving much better performance in every respect with only a 60 H. P. new-production air-cooled radial dual ignition engine with one-half the operating cost and twice the dependability, better vision, easier ship to handle, easier to repair, and one that fills 55% of the present demand for aircraft. Note its sturdy construction, study its complete streamlining and total lack of parasite resistance. Note its completeness in every detail, including airspeed.

Note the construction of corrugated duralumin wing, which is a patent feature and a decided step ahead in wing construction and is directly responsible for the wonderful performance and light weight and strength in this wonderful little ship. When you inspect the details, remember that every part of the New-Day Barling has undergone stress analysis, material tests, and structural tests. These parts have been proved more than the Department of Commerce requirements. Wing has been tested under all conditions to 125% of its design load without structural failure, ribs have been tested to 300% of design load without structural failure, and remember that the new principles used in this design have enabled us to build the lightest three-place airplane available in the world today, and we are presenting the New-Day Barling NB-3 for your careful and thorough consideration in design, performance, economy, structural strength, efficiency, and method of manufacture.

Fuselage and Welding Department, showing progressive line production of steel-tube fuselages which are tacked in one jig, removed to special jigs whereby they move forward to the lower end of building where various operations are completed by each group of men who have particular jobs to do. In this method quantity production is insured, allowing 100% inspection and eliminating all chances for inferior workmanship. All production is carried out on the above plan.

224 • AVIATION QUARTERLY • Spring 1989

AVIATION QUARTERLY

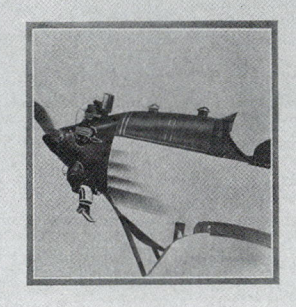

NOTE clean design and absence of wires and struts. Gas tank forms a part of top cowling and can be removed in five minutes' time.

SPECIAL heat-treated streamline chrome molybdenum tubing is used throughout; landing gear is attached to fuselage by four bolts and can be removed in fifteen minutes without disturbing the wing. Entire shock mechanism is inside of fuselage and is well streamlined, offering no resistance. It is very sturdy and has been designed for the hardest kinds of use, having school or student training especially in mind.

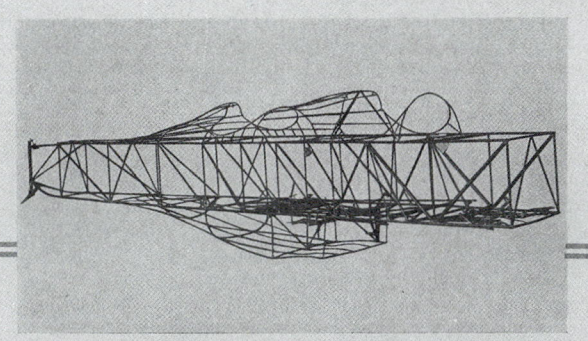

Fuselage is built entirely in jigs, using welded truss-type bracing. Seamless chrome molybdenum tubing is used throughout. Special jigs enable us to complete one fuselage every four hours. It is interesting to note that there is practically no wood used in the entire construction of the New-Day Barling.

ALL structural parts are made up of metal stampings which insure rapid production, uniform parts. Instead of using two beams as is used in conventional wing construction, the main part of wing is composed of one large Ubox made of corrugated dural and dural tubing, using steel tubes for drift bracing in the lower part. This patented structure is the lightest and strongest construction known today, due to the fact that all fittings are eliminated and stress or strain is not concentrated on any one part of the beam but is equally distributed over entire structure.

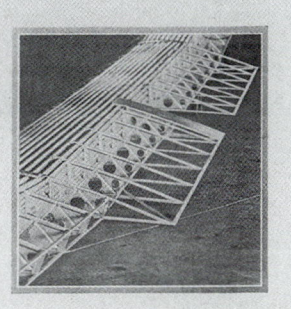

NOTE accessibility to lower part of wing for inspection purposes, and general construction for ease of repair. Complete wing can be attached or removed from fuselage by four bolts in fifteen minutes' time without disturbing landing gear. No rigging is necessary; however, provision has been made in both wing tips for wash-in or wash-out adjustment.

New-Day Barling patented wing, supporting 38 men or 6 1-2 times the total weight of the ship complete—one of the many grueling tests to which Barling parts have been submitted during the past 18 months' development period.

Spring 1989 • AVIATION QUARTERLY • 225

AVIATION QUARTERLY

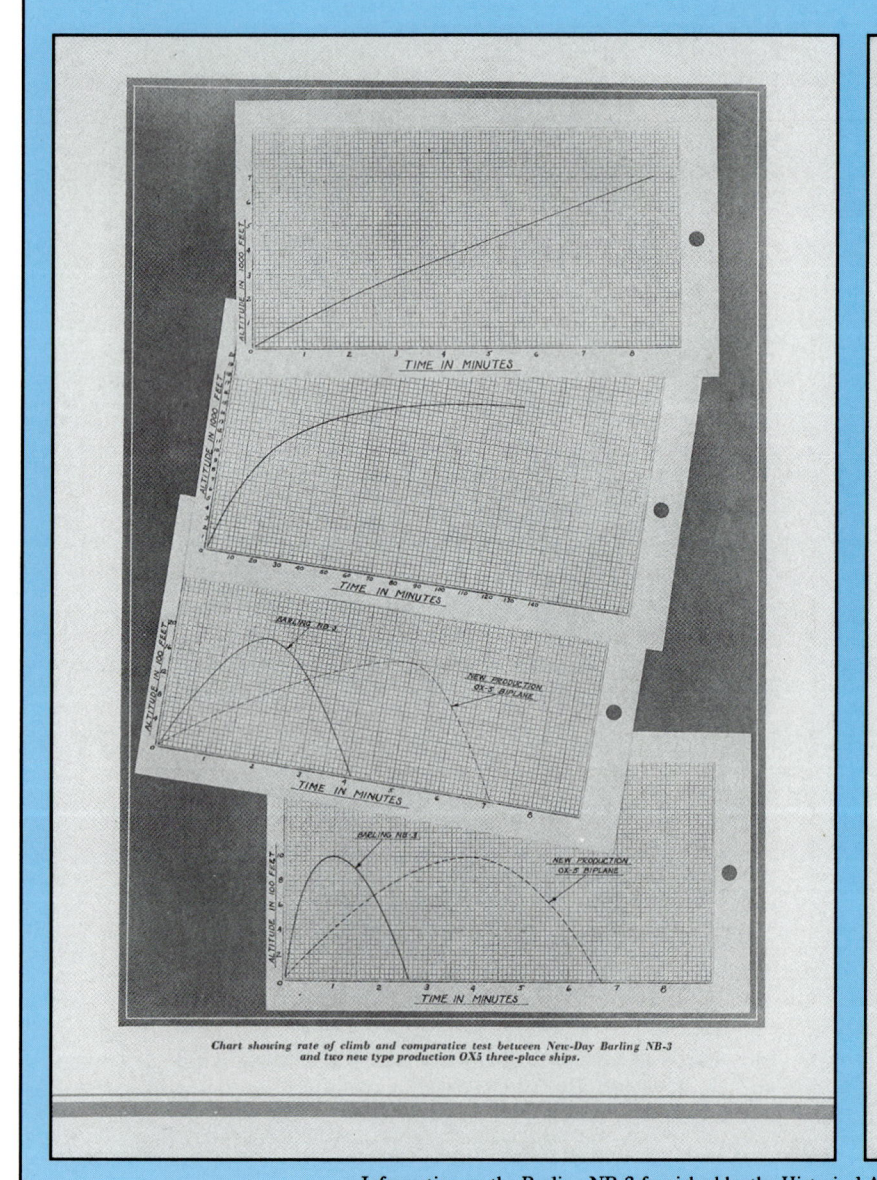

Chart showing rate of climb and comparative test between New-Day Barling NB-3 and two new type production OX5 three-place ships.

SPECIFICATIONS

Span	32 feet, 6 inches	Total Weight, Loaded	1,303 pounds
Chord	5 feet, 2 inches	Useful Load	613 pounds
Wing Area	159 1-2 square feet	Pilot	160 pounds
Area Tail Surface	25 square feet	Gas, 21 Gallons	126 pounds
Length	21 feet, 6 inches	Oil, 2 1-2 Gallons	17 pounds
Height	6 feet, 10 inches	Actual Pay Load	310 pounds
Dihedral	5 degrees	Wing Loading	8.3 pounds per square foot
Weight, Empty	690 pounds	Power Loading	20 1-2 pounds per H. P.

PERFORMANCE

PILOT ONLY		FULLY LOADED	
Take-off	4 seconds	Useful Load	613 pounds
Take-off Run	60 feet	Take-off	10 to 18 seconds
Run to Stop at Landing	50 feet	Take-off Run	100 yards
Climb to 1,200 feet	1 minute	Run to Stop at Landing	75 yards
Climb to 8,000 feet	8 minutes	Climb to 1,000 feet	65 seconds
Climb to 19,400 feet	124 minutes	Climb to 12,500 feet	45 minutes
Top Speed	105 miles per hour or over	Top Speed	105 miles per hour or over
Cruising Speed	87 miles per hour	Cruising Speed	87 miles per hour
Ceiling	21,000 feet	Service Ceiling	10,000 feet
		Absolute Ceiling	15,000 feet

Standard Color: Orange and Black   Price: $3,600 Complete

EQUIPMENT

Dual Control	Adjustable Pilot Seat	26 x 4 SS Tires	Wings Wired for Navigation Lights
First-Aid Kit	Fire Extinguisher	Carburetor Choke-dash Control	Altitude Adjustment
Altimeter	Tachometer	Oil Pressure Gauge	Oil Temperature Gauge

Air Speed
Dual Ignition Switch—Front and Rear   Gas Gauge   Tie-Down Hooks on Wings
LeBlond 60 H. P. Engine   Engine Tools   Wood Propeller and Cockpit Covers

MANUFACTURED BY

NICHOLAS-BEAZLEY AIRPLANE COMPANY, Inc.
MANUFACTURING DIVISION
MARSHALL, MISSOURI

Cable Address: NIBCO. Codes: Bentley's, Fifth Edition ABC, Western Union

Information on the Barling NB-3 furnished by the Historical Aviation Collection, University of Texas, at Richardson.

AVIATION QUARTERLY

VULTEE "VANGUARD" PURSUIT

MAKERS OF THE UNITED STATES AIR FORCE

8

Elwood R. Quesada: Tac Air Comes of Age

by John Schlight

Lt. Elwood R. Quesada

Only a few times in the history of military affairs have technological breakthroughs occurred that truly can be called revolutionary. The introduction of the stirrup in the fourth century, the longbow in the thirteenth, gun-powder in the fourteenth, and the airplane in our time qualify for that distinction. Each new invention, however, challenged comfortable ideas and forced a rethinking of accepted practices and doctrines.

Although among the first of the world's nations to purchase military aircraft, the United States after 1909 fell behind several other countries in taking advantage of the new weapon. Arriving late to World War I and buffeted afterwards by the twin obstacles of conservative military thinking and national economic plight, the airplane was not fully assimilated into America's military mainstream until the eve of World War II. Even then its role remained a matter of dispute, and it took that world conflagration to illustrate what military aircraft could do.

Through all of these fluctuations a relatively small group of men in the Army's aviation branch worked to discover and publicize the airplane's potential. These flyers shared some traits: a genuine enjoyment of the excitement and challenges posed by this new vehicle, a conviction that the airplane's potential was being stifled by institutional constraints, and a willingness— indeed an eagerness— to question standard solutions to problems. At the same time this group was far from monolithic. Often its members differed among themselves as to how airplanes should best be used. Bomber advocates believed fervently that not only their planes, but also their arguments in support of them, could bowl over any opposition. Others favored fighter planes, but not all for the same reasons. Some advanced the escort role of pursuit planes, others the interdiction capability of attack aircraft. But virtually all looked with disdain upon using airplanes to directly support ground forces.

Some of these early aviators fought principally with their pens, while others let their actions and personalities speak for them. Among the latter was Elwood R. "Pete" Quesada, who, in his twenty-six-year career, dealt in one way or another with most of the major issues associated with the growth of American air power.

The son of a Spanish businessman and an Irish-American mother, Quesada was born in 1904 in Washington, D.C., rather than in his parents' home in Spain. His mother preferred American doctors to deliver her children. He was, by his own characterization, "basically an immigrant." His entrance into the small aviation brotherhood in 1924 was as unorthodox as the way he achieved many of his later accomplishments. Quesada was quarterbacking the small University of Maryland football team that year when a moonlighting Air Service lieutenant who often refereed games recruited him to play

TAC AIR COMES OF AGE

for the team at the Army's flying school at Brooks Field in Texas in the fall. Enrolled there as a primary cadet with 150 other students, he took quickly to the World War I vintage Jennies that were used as trainers. By contrast the Brooks Field football team lost every game, and Quesada missed six weeks of flight training after he broke his leg on the gridiron. His strongly competitive nature led him to forego his Christmas leave to catch up under the tutelage of Lt. Nathan Twining, one of the school's instructors. The next spring he entered the advanced flying training school at nearby Kelly Field with the fourteen other successful primary graduates, including Thomas White, Earle Partridge and, for a while, Charles Lindbergh. At Kelly, Quesada learned to fly pursuit planes—the Sopwith SE-5 and the MB-3—and received his wings and a reserve commission as a second lieutenant.

In 1925 the Air Service was at its nadir. Still part of the Army, the fledgling aviation unit had to compete with other Army branches for funds and resources. It had only 880 officers and an equal number of planes, most of them obsolete. Flying officers were either recent West Point graduates or came from the other branches of the Army. Quesada was neither and returned to civilian life. While at Kelly he had played for the San Antonio Army Baseball Team in several exhibition games against the St. Louis Cardinals. Upon graduation he accepted an offer of $1,000 to sign on with the major league club. Quesada notes, with a scarcely concealed smile of irony, that the Army team's pitcher, Dizzy Dean, accepted $500 for the same offer. Quesada's string with the team lasted only a week. Aware of his vulnerability to certain pitches, he returned the money

Above: A flight of de Havilland DH-4s state side, post WWI. *(AQ Photo)*

The Curtiss Jenny was used as a trainer into the 20s. *(AQ Photo)*

MAKERS OF THE UNITED STATES AIR FORCE

to the club's manager, Branch Rickey, and ended his baseball career.

The outlook for the Air Service brightened the following year with congressional passage of the Army Air Corps Act. In addition to changing its name, the Army's air arm embarked, in 1927, on a five-year program to increase its numbers of officers and planes. A few inactive reserve officers, Quesada among them, were brought on active duty and given regular commissions. Working at the time as a criminal investigator for the Treasury Department in Detroit, he had felt the lure of the Army during his frequent visits to former classmates and familiar airplanes at nearby Selfridge Field.

His first assignment, to Bolling Field in Washington, provided important contacts and opened new vistas for the young lieutenant. Bolling at the time had a dozen different types of planes used by the Air Corps Chief, Maj. Gen. James E. Fechet, and members of his staff, including Maj. Carl A. Spaatz and Capt. Ira C. Eaker. Quesada's job as engineering officer was to keep the planes in good condition. He learned to fly all of them and quickly gained a reputation as a superb pilot and mechanic and as an energetic officer.

One Sunday in March 1928, while Quesada was riding a horse in Rock Creek Park, the Bolling commander drove up and informed him that Fechet wanted to see him immediately. A German plane, trying to be the first to fly the Atlantic from east to west, had gone down in Labrador, and the American government wanted the Air Corps to fly parts to the stricken plane. Two Loening amphibians, with Quesada at the controls of one and Eaker piloting the other, took off from Bolling. At the end of the first leg Quesada, with Fechet aboard, set the plane down on the Bay of Fundy only to have it stranded on the bay's sandy bottom when the tide went out. After digging holes in the sand under the plane's wheels to let down the landing gear, Quesada taxied up the beach and took off from land though he had come in as a flying boat. The mission succeeded, and in July, Fechet, impressed with Quesada's flying skill and ingenuity, made him his flying aide.

The planned five-year buildup of the Air Corps was slowed by the country's economic travails in the late twenties and early thirties. Air Corps officers still formed a relatively small and exclusive group, most of whom knew each other from their flying training days or subsequent assignments. Personal relationships played a large role in determining assignments and jobs. Much of the Air Corps' energy during these years was devoted to gaining public acceptance of the airplane as a versatile instrument for America's expanding society. This push for publicity had been behind the Air Corps' participation in the balloon expositions and speed races of the twenties and in Eaker's assignment as a pilot on a goodwill trip through Central and South America in 1927. It also partially explains the decision, in 1928, to see how long an airplane could stay aloft by refueling in the air. Quesada's close association with Fechet won him a berth as a pilot on the historical experiment. For nearly seven days in January 1929, along with Spaatz and Eaker, he helped fly a three-engined Fokker, called the Question Mark, as it orbited between San Diego and Los Angeles. Two modified Douglas C-1 transports, led by Capt. Ross Hoyt, served as "refuelers." Fuel was transferred through a hose, usually handled at the receiving end by Spaatz. Since there were no radios aboard, messages were written on blackboards, in whitewash on the fuselage of a plane that flew alongside, or attached to the end of the hose and supply line. Communication between a refueling crew member and his pilot was by means of a string tied to the pilot's arm. After eleven thousand miles, the Question Mark landed only because one of its engines began to falter. A generation later, while on an official visit to Vietnam years after his retirement, Quesada marveled at what this early flight had wrought as he witnessed wave after

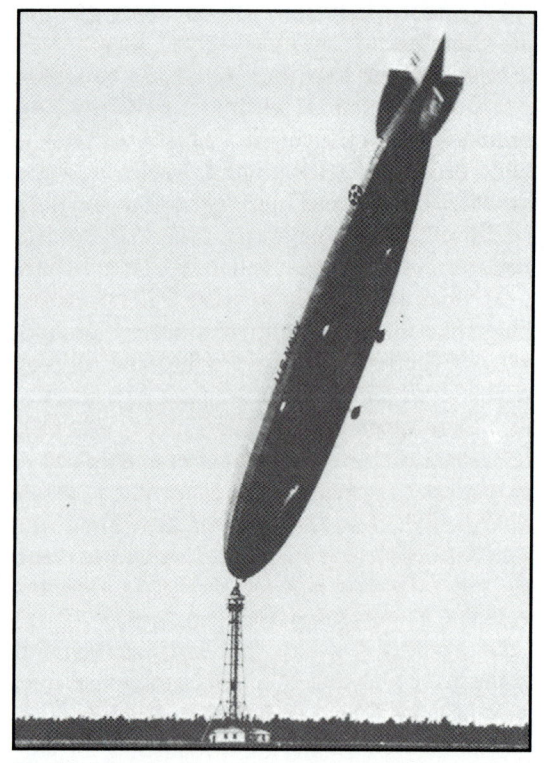

The ZR-3 *Los Angeles* was German-built and entered Navy service in 1924; retired in the early '30s. Improper handling resulted in this nosestand while anchored to mast in 1927. (USN) *Compliments Tab Books, Inc.*

TAC AIR COMES OF AGE

wave of American fighters and bombers refueling in mid-air for strikes against the North Vietnamese.

Following a two-year tour as an assistant military attaché in Cuba, Quesada returned to Washington as executive officer and flying aide to F. Trubee Davison, the Assistant Secretary of War for Air. In the summer of 1933, after Davison had left his post and became president of New York's Museum of Natural History, he and Quesada undertook a flying safari through Africa to gather animals for the museum.

The Air Corps' desire to gain a place in the sun was partially responsible for its unsuccessful experiment with carrying the mail the next winter. When in February President Roosevelt canceled as fraudulent the government's contracts with the commercial airlines, Quesada asked the Air Corps Chief, Maj. Gen. Benjamin Foulois, if Army flyers could handle the job. Foulois, according to Quesada, never one to say no, agreed. Between February and June 1934, Army pilots flew their transports, bombers, pursuits, and observation planes on missions for which neither they nor their equipment were prepared. Seat-of-the-pants flying, characteristic of much of Army aviation up until then, was unequal to the task. Most of the tactical planes lacked radios, gyros, artificial horizons, night-flying equipment, and even such rudimentary items as thermometers for detecting icing or cockpit lights for reading the instruments.

Quesada, by then a first lieutenant, was dispatched as chief pilot at Newark Airport in New Jersey, from where he flew round trips every other night to Cleveland. Piloting a Curtiss Condor, one of the Army's newest planes with navigational equipment

Participants in *Question Mark* refueling, from l. to r.: Capt. Ross G. Hoyt (pilot of refueling plane); Capt. Ira C. Eaker; Maj. Gen. James E. Fechet (Chief of the Air Corps, not a crew member); Maj. Carl Spaatz; Lt. Elwood Quesada; and MSgt. R. W. Hooee. *(Photo: USAF)*

aboard, he had little trouble. On half of his westward flights, however, headwinds forced him to land at the same small, private airfield in western Pennsylvania to refuel. After several weeks the field's manager, weary of rising at two in the morning to pump gasoline in subzero and snowy weather, left the key in a prearranged hiding place so that Quesada could help himself. The return runs, with the wind at his back, required no stops. One of them he flew in the record time of one hour and twenty-seven minutes. Pilots on the eighteen other runs were not so fortunate. Before the airlines resumed carrying the mail in June, sixty-six Air Corps planes had crashed and twelve pilots had died.

Quesada's spreading reputation as a diplomat and superior pilot netted him several more assignments as flying aide and executive officer to important government figures. After the airmail experience, in the summer of 1934, he was assigned to Hugh Johnson, the administrator of Roosevelt's National Recovery Administration (NRA). In addition to flying Johnson to all parts of

MAKERS OF THE UNITED STATES AIR FORCE

the country, Quesada acted as a research assistant, helping him with congressional testimony and checking the accuracy of his speeches. With the demise of the NRA, Quesada became, briefly, an assistant to Secretary of War George Dern. For a short time after that he was an aide to Brig. Gen. George Marshall at Fort Benning, Georgia.

Quesada later acknowledged the value to him of these years as executive to the mighty. While the contacts he made were important, even more valuable in later years were the useful skills he developed in understanding the mind-set of military and civilian leaders, in sharpening a keen negotiating ability, and in being able to plan from a perspective that placed the Air Corps against a larger background. This latter trait, in particular, set him aside from many of his contemporaries. Continued service as an aide also prompted Quesada to the belief that his career was taking on a lopsided appearance, and he welcomed a fresh assignment.

Early in 1935, Quesada joined Col. Frank M. Andrews as the new GHQ Air Force was getting underway. Billy Mitchell's vitriolic campaign in the twenties for an air force separate from the Army had alienated many Army officials. Successive Air Corps Chiefs, Fechet and Foulois, through patience and diplomacy, had calmed the fears of many among the opposition, and by 1934 some Army leaders were beginning to see merit in a separate branch of bombers, pursuit, and attack planes which, still part of the Army, would provide a semblance of independence. The result was a new task force, the GHQ Air Force, whose creation, it was hoped, would deflate the Air Corps' resurgent move toward divorce.

On the first day of 1935, Andrews was appointed Commander of the GHQ Air Force with the job of organizing the force and activating it by March. At first, with only Quesada at his elbow in the State, War, and Navy Building in Washington, Andrews organized the GHQ Air Force and selected the people he wanted on his staff while Quesada handled the paperwork. When the outfit moved to Langley Field, Virginia, several months later, Quesada went with as commander of the headquarters squadron which involved, among other things, "kicking people out of barracks so the GHQ could come in."

The move to Virginia went smoothly, and Andrews, now a brigadier general, rewarded Quesada with an assignment he wanted, to the Air Corps Tactical School at Maxwell Field, Alabama. At that time the school was a center of ferment. In contrast to the War Department's defensive view of aviation, the Tactical Schoiol taught that the offensive mission of strategic bombing was more decisive than supporting ground forces or defending the coastline. Corollary to this notion, and possibly partially responsible for it, was the view widely held among aviators than an independent bombing mission would never be accepted as long as the Air Corps remained under the thumb of the Army. Quesada came to share these views with what he later called the "agitators," noting that "agitation is how you get things done very often."

His year at Maxwell was followed by a year at the Army's Command and General Staff School at Fort Leavenworth, Kansas, and then to his first operational assignment as a flight commander in a bombardment squadron at Mitchel Field, Long Island. In this job less than a year, he was sent, in the summer of 1938, to Argentina as a technical advisor to the military. The Argentinian Air Force, which at the time had 150 American bombers and fighters, was trying to pattern itself after the American Air Corps. Chosen for his mechanical and organizational ability, Quesada was assigned to help them. He and four other American aviators assisted in installing maintenance and supply systems and a method of instruction for blind flying. The Argentinians had received a license to build their own aircraft engines and had opened a factory at Cordoba to produce them. Their lack of organizational skill, however, was making it costly for the government, both economically and politically. Being the only bachelor among the Americans, Quesada was sent to Cordoba, "where there was nothing but a wonderful climate," to install a system of quality control and inspection.

Curtiss D-12 engine with 400 to 460 hp. *(Photo: USAF)*.

TAC AIR COMES OF AGE

"READY!" *(Official U. S. Navy Photograph)*

The night before Quesada left Buenos Aires for the States in September 1940, some American naval officers threw a party for him. At the height of the festivities he promised to relieve them of an unwanted Grumman amphibian plane by flying it home. With only a few five-gallon tins of gasoline, a screwdriver, a pair of diagonals, and some safety wire on board, he took off the next morning, flew over the Andes to the Pacific, and up the coast to Panama. His unannounced landing at the tightly guarded American naval air station there caused an uproar. As he climbed from his plane the airdrome officer asked who he was. Unable to locate a Captain Quesada in the Navy's register of officers, the AO summoned the station's commanding admiral. It took some time for the impish Quesada to clear up the suspicious phenomenon of an Army officer with a Spanish name flying a U.S. Navy plane from Argentina. Several days later he completed the flight to Norfolk, Virginia, becoming the first aviator to fly that rout solo.

* * * * *

By the time Quesada returned from Argentina, the nations of Europe had been warring for a year, England was successfully weathering Hitler's bid to cow it from the air, and Maj. Gen. Henry H. Arnold was Chief of the Air Corps. As a result of his overseas experience, Quesada became Arnold's foreign liaison chief, working closely with the British embassy to supply them with information about Air Corps planes and equipment. When Arnold went to England in

MAKERS OF THE UNITED STATES AIR FORCE

April 1941, Quesada accompanied him. The Lend-Lease Act was only a month old, and one of Arnold's purposes was to make arrangements with the British for flying American planes to England. As the Chief's aide, Quesada did the spadework for the trip, negotiating the itinerary, preparing the issue books, and memorizing scores of facts for the discussions. While in London the two witnessed the German bombing of the city. In visits to Royal Air Force bomber and fighter bases Quesada was impressed with the courage and determination, less so with the quantity of equipment, of the British flyers. Agreements were reached on Air Corps-RAF cooperation. Out of this visit was born the Ferrying Command, predecessor of the Air Force's later worldwide military airlift organizations.

Once back in Washington it fell to Quesada to set in motion many of the programs that flowed from the agreement—arranging for people to administer the Air Corps' portion of the Lend-Lease plan, transferring aluminum to England, and training British pilots in Air Corps schools. For his trouble, Arnold, now Commanding General of the newly designated Army Air Forces, rewarded the major with a promotion and his much-sought-after command of the 33d Fighter Group of P-40s at Mitchel Field.

When the United States entered the war in December, the Army created the Eastern Theater of Operations under Gen. Hugh Drum to defend the east coast against German submarines and, although not seriously expected, German air attacks or landings. The First Air Force became part of this defense force. In August 1942 Quesada, by now a temporary colonel, turned over the 33d to one of his squadron commanders,

234 • AVIATION QUARTERLY • Spring 1989

TAC AIR COMES OF AGE

William Momyer, and took charge of the I Fighter Command. His job, in coordination with the Army's antiaircraft artillery, was to train P-40 pilots to intercept enemy planes, particularly at night. No sooner had he settled in his Philadelphia office than he landed in a jurisdictional dispute with the Army's artillery commander, Brig. Gen. Sanderford Jarman, over who was responsible for identifying planes flying in the area. It was the Army's practice to identify airborne planes at night by playing searchlights on them. After one of his pilots, blinded by the lights, crashed while trying to land, Quesada issued a written order banning the searchlights. He included in the order an obiter dictum that, since German planes could not much more than cross the twenty-two mile English Channel from their home bases, it was to be assumed that there were none within a thousand miles of the United States. All planes flying over the U.S., therefore, were friendly. A sharp verbal exchange ensued between the colonel and the general, after which Jarman preferred charges and sought to courts martial Quesada. The Army Chief of Staff, General Marshall, sent his inspector general and General Arnold to look into the matter. Both reported that, while Quesada had been impulsive, the Army practice was wrong and the courts martial should be dropped. Marshall, more than any of the top Army leaders, appreciated air power's potential and the need to bring bright, young, energetic officers into the general officer ranks. The lights were turned off, the charges dropped, and Quesada was promoted to brigadier general and returned to Mitchel Field to activate the 1st Air Defense Wing.

Early in 1943 Quesada took his defense wing to North Africa, where it joined the XII Fighter Command in defending the Allied forces against air and submarine attacks, and in protecting friendly shipping and attacking enemy convoys in the Mediterranean. Within a month he was commanding the XII Fighter Command.

Shortly before he arrived in Africa, at the Casablanca Conference in January 1943, the decision was made to unite British and American forces into functional, combined commands—all bombers in one command, fighters in another, air defense planes and equipment in a third, and training in a fourth. It was hoped, correctly, that this would help stop bickering among the Allies and, more importantly, end the piecemeal dissipation of air resources that had been taking place. Quesada was present at a meeting in Tripoli, attended by Spaatz and the other American and British air leaders, to decide who would head each of the four new combined air forces. He remained uncharacteristically silent for a long time as the British argued forcefully for control of all the commands since they had had more experience in the war. Finally, unable to contain himself longer, Quesada reminded the airmen that this experience they claimed included Dunkirk, Singapore, Crete, Greece, and a host of other British Army losses. Spaatz leaned over, tapped Quesada on the knee,

Curtiss P-40 aircraft was one of the basic fighter interceptors of the U.S. Army Air Forces relied on during the time leading up to U.S. entry into WWII. *(AQ Photo)*

MAKERS OF THE UNITED STATES AIR FORCE

P-51B drawing by Walter C. Fink III, A 737 Captain for United Air Lines.

and said, "Take it easy, Pete." The British airmen, principally Air Vice Marshal Sir Arthur Coningham, who harbored an undisguised dislike for some of the early leaders of the British Army, got the point and agreed to Quesada's suggestion that they choose commanders according to the predominance of force in each command. The British were given control of the tactical and air defense air forces, while Brig. Gens. Jimmy Doolittle and John K. Cannon took over the bomber and training commands. Quesada, whose XII Fighter Command was absorbed into the new Northwest Africa Coastal Defense Force under Sir Hugh Pugh Lloyd, donned a second hat as deputy commander of that defense force.

One of the more important results of this meeting was the friendship that sprung up between Quesada and Coningham. The air marshal had led the British tactical air forces at El Alamein the preceding winter. By force of personality he had convinced Gen. Bernard Montgomery that he could get the most out of his planes, not by using them as artillery, but by letting them wrest the air from the Luftwaffe and attack German airfields and lines of communication behind the front. Quesada and the other American tactical flyers had been advocating such independent use of tactical air power but had been unable to win their point as convincingly as the air marshal. Through his acquaintanceship with Coningham, Quesada would become one of the major conduits through which tactical air doctrine and practice would flow from Africa to continental Europe and eventually to the United States Air Force.

With the Germans cleared from Africa and the Allied invasion of Sicily and Italy well established, most of the major leaders of the African campaign moved to England in October to prepare for the invasion of the continent. Quesada went along with Eisenhower, Tedder, Bradley, Patton, Coningham, and, later, Doolittle, in transferring the African system of combined organizations to England. Lacking a long American tradition of tactical air power, Quesada and the other tactical planners fell back on what they had learned from both British and German air operations so far in the war, much of it formalized in Field Manual 100-20, which had been written and approved during the North African campaign. They borrowed their organizatin from the British, inspired by Coningham's insistence that tactical aircraft cooperate with the ground forces as an independent force. The American IX Fighter Command was set up as a coequal of the numbered American army, to work closely with it. An analysis of the Luftwaffe campaigns over Poland, France, England, and Africa led them to conclude that, while the German Air Force had achieved some dra-

236 • AVIATION QUARTERLY • Spring 1989

TAC AIR COMES OF AGE

matic local successes, its failure to gain control of the air at the outset doomed it to eventual defeat. Furthermore, by being tied too closely to the German armies, the Luftwaffe lacked the flexibility it needed for ultimate success.

Allied preparation for the invasion was divided into two phases. At first, Allied planes would gain control of the air over France and the Low Countries and destroy enough of Germany's industrial base to make a landing possible. In the forefront of this phase were the heavy bombers of the Eighth Air Force and British Bomber Command, supplemented, as they became operational, by the medium bombers and fighters of the newly created Ninth Air Force. Immediately prior to the invasion, tactical air units would participate in isolating the landing areas by knocking out roads, bridges, and rail lines. In the second phase, during and after the landings, the Ninth Air Force and the British Tactical Air Force would back up the ground assault.

The Ninth Air Force, commanded by Maj. Gen. Lewis H. Brereton, was being built from a few remnants of the old Ninth from Africa, the medium bombers transferred from the Eighth in England, new units arriving from the States, and unattached airmen already on the island. Units for Quesada's IX Fighter Command began arriving from the States in November 1943 at the rate of one or two a week. Thanks to prior planning and Quesada's energy, most of these fighters were flying missions within three days of their arrival. By December he had built his command to two wings, each with four eighty-plane groups of P-38s, P-47s, and the newly introduced P-51 Mustangs. One of these outfits was a reconnaissance group. At the end of February, the command had grown to four wings of nine groups. To get ready to support the two American armies, the First and the Third, after the landings, Quesada's fighter command, swollen to eighteen groups by May, was split into two tactical air commands, the IX Tactical Air Command (TAC),

MAKERS OF THE UNITED STATES AIR FORCE

which Quesada commanded directly, and the XIX TAC, led by Brig. Gen. O. P. Weyland. Quesada was dual-hatted, commanding both the parent IX Fighter Command and one of its two tactical air commands. All told he was responsible for more than 1,500 fighter planes.

Integrating and training the new fighter groups was hectic. Months of training were compressed into weeks as the fighter command prepared for the invasion. Quesada was determined to indoctrinate his incoming flyers with the interdiction techniques he had witnessed in Africa. He brought generals and other specialists who had flown in the African campaign to lecture to the aviators. He sent officers to Italy to observe the methods of cooperating with the ground forces being used by the tactical planes of the AAF's Twelfth Air Force, and then brought them back to assist his training program. Reflecting his understanding of human nature, and to stimulate his flyers' interest, Quesada labeled these activities "combat drill," rather than the unpopular "training." He set up bombing courses and ran the pilots through drills in dive and glide bombing, low-level attacks, night flying, low-level navigation, patrol convoys, and smoke laying. As part of his program to convert fighters into fighter-bombers, he directed his pilots on practice missions against bridges, locomotives, trucks, and tanks at the British Millfield

The deadly de Havilland *Mosquito* fighter-bomber. *(AQ photo)*

School training center. He even devised a means of hanging a pair of 1,000-pound bombs on his P-47s. Unhappy with the second-class status accorded reconnaissance in Africa, and deeming his one reconnaissance group insufficient for such a large command, he insisted that his fighter pilots become proficient in reconnaissance. Little joint training with the ground forces took place, however. The prewar attitude that close air support of ground forces was not a priority air mission still prevailed among flyers at all levels.

Improvisation abounded. Quesada, sensitive from his African experience to the critical importance of communications for such a wide-ranging operation, helped to devise a telephone system that was to prove invaluable after the invasion. One day a young draftee who had been an AT&T technician before the war demonstrated for him an FM transmitter and receiver he had bought in a surplus store in New York City. America's conversion from AM to FM had been halted at the outset of the war, and much of the unused equipment had been sold as surplus. Quesada checked out the equipment by setting up the transmitter at his headquarters at Middle Wallop and the receiver at Land's End, the same distance from his headquarters as was Normandy. The excellent static-free reception convinced him to take some sets with him to the continent. Producing $600 from his pocket, he sent the technician back

238 • AVIATION QUARTERLY • Spring 1989

TAC AIR COMES OF AGE

to New York to find as many sets as he could and bring them back to England. The equipment was loaded on a landing craft and used on D-day.

While touring an early-warning radar site in southern England, Quesada put his air defense experience to good use. The radar sets were being used defensively to detect enemy planes. As he was looking at the screen at one of the sites, a series of blips appeared over Brest. Since the normal flight path of returning American bombers was over Dover, it had been assumed that the radar blips were German planes. At the same time conversations could be heard from a radio receiver elsewhere in the room between the crews of an American bomber force that was lost. Quesada had the operators install the right crystals to talk to the Americans. When he told the pilots to turn right, the images on the screen followed his instructions. He realized immediately that this defensive equipment could serve an offensive purpose. By connecting a Norden bombsight, upside down and backwards, to the radar, he helped create a ground control system his fighters later used to great advantage on the continent during bad weather. This same principle was later used by the Strategic Air Command to score practice bombing missions, and in Vietnam to control fighter and bomber strikes.

These "combat drills" were sandwiched in between operational missions. The fighters of the Ninth flew their first missions early in December 1943, escorting the bombers of the Eighth and Ninth Air Forces over France and the Low Countries. In mid-month, P-51s, with Quesada flying one of them, set a distance record escorting B-17s to Kiel, a round-trip distance of almost a thousand miles. On many of these missions, however, the Mustangs developed firing problems. The planes' guns started jamming after sharp turns, leaving the pilots at a disadvantage against German fighters. Quesada knew instinctively that these hundreds of young, untested aviators would fly better if they trusted their leaders to solve their problems. After examining the post-strike photography, he drove over to Brereton's headquarters where he complained vociferously about the malfunctioning guns "his kids" had to use. Brereton's response was to put in a call to General Arnold. When the chief came on the phone, Brereton handed the instrument to Quesada saying, "Alright, Pete, tell him what you told me." Never one to back away from a challenge, Quesada told Arnold about the mess and how his pilots deserved better. Arnold promised over the phone that he would have it cleared up. That night a group of technicians was on its way from Wright Field, and within days the problem was solved.

Attacking problems head-on became a Quesada hallmark. When the Mustangs experienced a rash of problems with fouled spark plugs on return trips from escort missions, many pilots were forced to ditch in the North Sea. Some were lost because, in the excitement of their preflight check, they had failed to hook their dinghies onto their flight suits. While setting technicians to solve the spark plug problem, Quesada also took immediate steps to make the crew chiefs responsible, through written statements, to see that the dinghies were attached to the pilots' harnesses before takeoffs.

In December 1943, Quesada's fighters became part of the combined British-American tactical air force, the Allied Expeditionary Air Force, commanded by Air Marshal Sir Trafford Leigh-Mallory. Originally Quesada was supposed to get all the Mustangs while Eighth Air Force got the P-38s and P-47s. When the longer-range Mustangs proved to be better air-to-air fighters, Eaker, then commanding the Eighth, implored Brereton to let some of them come to his command. Quesada, knowing that the Thunderbolts and Lightnings were better for interdiction and close air support missions, persuaded Brereton to let some of them go.

The primary mission of Quesada's planes was to help gain control of the air over France before the invasion by defeating the German Air Force. Until April 1944 they did this, in between training, by escorting the bombers, and by tempting German fighters into the air and destroying them. As a second priority the fighters hit installations along the French coast that the Germans were build-

Spring 1989 • AVIATION QUARTERLY • 239

MAKERS OF THE UNITED STATES AIR FORCE

ing to launch their buzz bombs. Direct attacks on air fields and industrial complexes were last on the list. By March, many of the pilots were becoming impatient with escorting bombers, and anxious to fly more interdiction missions. The following month, for the first time, the number of dive bomber missions outnumbered escort flights.

During these preinvasion months, Quesada flew along with his pilots. While no law prevented a general from flying, it was frowned upon. But Quesada insisted that he had to know what was going on and that he could make better decisions based on experience than on operational reports. He adopted the attitude that if others could go he should be allowed to, especially since he was "just as expendable and damn near as young as his pilots. Also, he conceded, he "didn't want those little jerks to think I couldn't fly as well as they could." His only hesitation stemmed from concern that the other pilots would feel they might have to devote a lot of attention to protecting "the old man."

By April the Ninth began to move away from escorting bombers to the priority missions of tactical aircraft—gaining control of the air and isolating the battlefield. Units moved into Hampshire along the southern coast, and Quesada shifted his headquarters to the western edge of London at Uxbridge. Hundreds of daily flights, in concert with the bombers, wiped out the German airfields in France and bombed railway centers, marshaling yards, bridges, rolling stock, and coastal batteries. This phase of the campaign was aimed at weakening the enemy to the point where he would be unable to match the Allied rate of buildup around the beachhead once the invasion took place. To mask the landing site, the Ninth Air Force divided its efforts between the Cherbourg Peninsula and the Pas-de-Calais area, 160 miles to the north, where the enemy was expecting the invasion.

At a meeting of the top Allied military commanders several days before D-day, all eyes turned toward Quesada when he predicted that the landing would go unopposed by the German Air Force. "How can you be sure?" asked Churchill, who was presiding. The uninhibited general responded confidently that Allied fighters had met decreasing resistance from German planes over the past six weeks, that control of the air was an established fact, and that it would stay that way.

This intuition proved correct when only 750 German planes took to the air on D-day, most of them far from the beachhead. The landing went smoothly with Quesada's fighters covering the beaches, escorting sea and air convoys, and hitting coastal batteries, enemy troops, and bridges behind the landing sites. Only on Omaha Beach was the landing stalled. By shifting several of his groups from air-to-air missions to a full day of strikes against German artillery, he helped to remove the obstacle.

Leaving his XIX Tactical Air Command behind in England to prepare under Weyland for its later action with Patton's Third Army, Quesada flew his P-38 to France the following day, landing "with one wing over a cliff." Near the beachhead he set up his command

P-51 in British markings. (Photo: USAF)

TAC AIR COMES OF AGE

P51Bs shown attacking Nazi held installation on European continent in 1943, thru Artist's eyes.

post alongside that of Gen. Omar Bradley, with whose First Army he would be working.

Communication equipment and control radars were quickly installed, and the engineers began building a dozen airfields so the fighters could move over from England. By the end of June, nine of the fields were in use, and seven groups of fighters had moved over premanently from the island. Several days after the landing, General Eisenhower arrived at the command post for a visit. When he was ready to return to England he asked Bradley to radio ahead his arrival time. Quesada offered instead to call by radio telephone, and within minutes the supreme commander was talking to his headquarters in England over Quesada's static-free hook-up. Eisenhower was surprised to learn of Quesada's superb communications which surpassed those of the ground forces.

Several weeks later, during a subsequent visit, Eisenhower expressed interest in accompanying Quesada on a fighter mission he was planning to fly over Paris. The Germans had moved a large number of airplanes into ten fields around the French capital, and Quesada agreed to take the commander along to have a look. At the short, unfinished airfield, which had steel planking for a runway, Eisenhower climbed into a makeshift back seat of a P-51 from which the 70-gallon fuel tank had been removed. With Quesada at the controls, they took off and linked up with the other planes of the flight. Quesada prudently decided against going to Paris, and they flew instead over the battlefield about fifty miles south of the field. Eisenhower was full of questions and impressed with the formation flying, breakaways, and the communications. Landing back at the field, they were spattered with mud, evoking

MAKERS OF THE UNITED STATES AIR FORCE

from Eisenhower the comment that his friends had misinformed him by telling him that airmen lived in hotels. Although this excursion brought down on Eisenhower a strong rebuke from Marshall, and Spaatz slapped Quesada's wrist, the IX TAC Commander felt it was worthwhile in cementing relations and giving Eisenhower firsthand experience with the air forces.

Except during the capture of the city of Cherbourg late in June, Quesada's fighters concentrated until late July on interdiction missions. Close air support was not entirely absent, however. The Normandy countryside proved a surprise for both Bradley and Quesada. High hedges that separated the numerous fields provided a natural defense for the Germans, making it more necessary than anticipated for the fighters to strike close to the American lines. But basically the American infantry fought its way down the peninsula while the fighters struck targets outside their view. One job for the fighters was to make it as difficult as possible for the German divisions, which had been poised around Calais, to move westward when they realized that the only landing was to be in Normandy. Quesada's planes, along with the Royal Air Force, hit these troops day after day, and they arrived in Normandy not as unified divisions but as disorganized mobs of tanks. On one occasion, learning of a possible meeting between Rommel and some high German officials, Quesada led a flight of P-38s and destroyed a number of vehicles and buildings at the suspected site. The results were never learned.

When the fighting reached the base of the Cherbourge peninsula and the Allied troops encountered stiff German resistance around St. Lô, this situatin began to change. Flying

TAC AIR COMES OF AGE

Luftwaffe pilots called them "forked-tail" devils. (Photo: Lockheed Aircraft Corp.)

Lockheed's P-38 Lightnings

back to England, Quesada and Bradley conferred with Spaatz, Leigh-Mallory, and Doolittle in planning a carpet bombing strike against the enemy to soften them up for the American attack. Back in Normandy a few days before the attack, Quesada suggest to Bradley that the tanks could best exploit the paralysis caused by the bombing if they concentrated their attack into several columns rather than across a broad front. Bradley agreed, and Quesada promised to keep a continuous cover of fighters over the head of each column to warn the tank commanders of hidden enemy forces and to respond to their calls for assistance. He also told Bradley that if the Army would send a tank over to his headquarters he would install Air Force radios in it and, if the experiment worked, would assign pilots to the lead tanks to operate the radios and advise the ground commander what the airmen saw. In this way the pilots, trained to spot and describe targets as they would appear from the air, could talk in airmen's terms to their counterparts flying above. Bradley agreed and ordered his ordnance people to send a tank to IX TAC Headquarters. The soldiers thought the general had made a mistake and sent it, instead, to the 9th Infantry Division. Quesada, becoming impatient, called Bradley, who then had it sent to the correct place. When it arrived the Air Force guards would not let it in, seeing no need for a tank. The situation was cleared up, the radio installed, and the experiment worked. Quesada was in on the creation of an element of air control parties that would form the nerve center for later tactical air efforts in Korea and Vietnam.

The subsequent breakthrough at St. Lô was in many ways a turning point not only for tactical air doctrine but for Quesada as well. Having shared until then the flyer's almost universal aversion to working too closely with ground troops, he underwent what was almost a battlefield conversion in coming to appreciate at close range the necessity of cooperation.

A large part of Quesada's success arose from the mutual understanding and confidence that had developed between him and Bradley. Suspicion between other ground and air commanders had not entirely evaporated and was kindled by inevitable mistakes on both sides: antiaircraft artillery occasionally shooting down American planes and the fighters at times inadvertently

MAKERS OF THE UNITED STATES AIR FORCE

bombing friendly troops. The two generals worked together to dispel these antagonisms and to make their own commanders appreciate the conditions under which the others worked. At breakfast one morning, for instance, Bradley showed Quesada a message from V Corps complaining that a planned major offensive had been canceled the day before because of a heavy unintercepted German air attack. Enraged that he had not heard of it, Quesada checked with his operations officer only to learn that two German planes had flown over a regimental headquarters destroying one truck and injuring one man. Knowing that actions spoke louder than words with Bradley, Quesada persuaded him that the two of them should investigate the incident personally. Picking up the corps and division commanders on the way, the four generals confronted the regimental commander, who pointed to the charred remains of a truck and introduced them to the regimental cook who limped up with a shrapnel wound in his rump. Quesada then pulled from his pocket a list he had compiled earlier and read off the Allied air action of the previous day—1,000 bombers had dumped 4,000 tons on Germany, and on the way home 600 fighters had straffed everything in sight. "But our whole army," he concluded, "was stopped by two planes that dropped no bombs, set a half-track on fire, and shot a cook in the ass! If air power is as effective on the Germans as it seems to be on us, why aren't we in Berlin?" Bradley and Quesada rode home in silence. The following day the Army commander sent a strongly worded letter to all his commanders outlining the incident and telling them that they must be prepared for an occasional air attack. Bradley was aware that Quesada had staged the inspection, and Quesada knew that what he had done was a better way of getting his point across than simply telling his counterpart.

Quesada's intimate working relationship with the Army, however, was not without its price. He was criticized by some airmen who felt that his support of the infantry was abetting those who thought the Army should have its own air force. Quesada waved such objections aside. He was where the fighting was, and his job was to help in any way he could. He could't "just sit there and say Hell, no!"

On the march to Avranches, which fell near the end of July, tanks and planes cooperated in numerous ways. On one occasion a Sherman tank, surrounded ty thirteen Tigers, was saved when a squadron of P-47s scattered the enemy's vehicles. Often the Thunderbolts, in response to Army appeals, cleared roads of enemy tanks lying in ambush. Once, when the radios went dead, the tanks shelled a railway station as a signal for the dive-bombers to attack the station. At another time the tanks used their tracer bullets to mark targets for Quesada's fighters.

Maj. Gen. Quesada seated in the cockpit of a Lockheed P-38. (Photo: USAF)

244 • AVIATION QUARTERLY • Spring 1989

TAC AIR COMES OF AGE

The airplanes of IX TAC were tied to the fast-moving army by Quesada's fighter control radar. Set up in tents and constantly moved eastward to keep pace with the ground forces, the radar scanned the entire battlefield and directed the fighters as they flew between 1,300 and 1,800 sorties each day. When the planes flew cover for the tanks, the control tent turned direction of them over to the pilot in the lead tank of each column.

It was the First Army, now under the command of Gen. Courtney Hodges, that received the initial jolt of the German counteroffensive in the Ardennes shortly before Christmas. From his headquarters at Verviers, just north of this Battle of the Bulge, Quesada directed his fighters, and those of the British 2d Tactical Air Force which Coningham had turned over to him, against the advancing German tank columns. Bad weather hampered operations at first. On the 18th, as the Germans moved through a total ground fog, Quesada asked for volunteers from the reconnaissance group to fly through the mist and check on the German advance. Two Mustangs, flying through the fog a hundred feet off the ground, discovered sixty tanks and armored vehicles moving toward Stavelot. Ninth fighter bombers spent the rest of the day destroying most of them. After the weather cleared on the 23d, Quesada's fighters, along with those of Weyland's XIX TAC, helped to halt the Germans by cratering and cutting rail and road lines, blocking chokepoints and narrow passes, and destroying many tanks and vehicles.

None of Quesada's bases were overrun during the German counteroffensive, although at one point ground attacks against some of them seemed imminent. To prepare for the expected assault Quesada ordered the antiaircraft artillery moved out onto the nearby roads and pointed horizontally down the route along which the enemy was expected. The tanks never appeared. The Germans seldom hit the fields from the air. One exception occurred, however, on the first day of 1945. As a companion to their ground attack, the Germans planned a massive strike against Allied airfields by an armada of Luftwaffe planes. Alerted to the enemy's intention by the Ultra system, which was reading German code traffic, Quesada ordered pilots stationed at the antiaircraft positions to identify the planes for the gunners. When the attacks came, not one Allied plane was lost.

Quesada's converted radar sets came into their own during the Battle of the Bulge. Pilots flying over the snow-covered terrain were having trouble distinguishing enemy tanks and vehicles from friendly ones. Quesada moved two of the radars close to the action, and the radar operators, who knew where the front line was, advised the pilots when to strike and when to hold their fire. The radar operators knew better than the pilots whether they were or were not on the German side of the line. At the same time the Ninth's long-range radars were instrumental in guiding dozens of stricken B-17s, returning from bombing runs over Germany, safely onto local airfields.

Having eliminated the German salient, the Allied armies pushed on to the Rhine at Cologne. The biggest obstacle in their way, the city of Duren, was heavily bombed by RAF and Eighth Air Force bombers and Quesada's fighter bombers. When the advance ground units discovered the bridge at Remagen intact, Quesada assigned his fighters to fly patrols over it for four days to ward off any German attempts to destroy it as Allied soldiers streamed across. He stationed one of his ground controllers high up on the bridge to direct the air battle if one developed over the structure. Neither of these actions, however, proved necessary.

Once across the Rhine, Quesada's planes assisted the American armies in trapping more than 100,000 German soldiers around Paderborn. Discovering a hornet's nest of German airfields there, his fighters destroyed hundreds of Nazi planes, including jets. For all practical purposes the German Army was defeated, and the Allies met little further resistance during their subsequent march to the Elbe.

Throughout the sweep across France, the Low Countries, and Germany, Quesada earned a reputation for dash, imagination, and above all, leadership. Sensitive to the concerns of his men and of their parents, relatives, and well-wishers back home and in

P47D ready for another mission in 1944 or late in the war. (Photo: USAF)

MAKERS OF THE UNITED STATES AIR FORCE

England, he kept up a voluminous correspondence with them, allaying their fears and keeping them abreast of the war's progress. In a letter to Quesada's mother, his aide noted that the general had become a "star" and a hero to many concerned with America's success. "Quesada was a peach to work with," wrote Bradley to Arnold in September, "because he was not only willing to try everything that would help us, but he inspired his whole command with this desire."

* * * * *

With the surrender of Germany accomplished and the defeat of Japan assured, AAF leaders resumed their prewar campaign for a separate air force. Newly elected President Harry Truman was known to favor a reorganization of the military structure. In an attempt to gain direct access to him, Spaatz and Arnold tried to have Quesada assigned as his military aide. The general returned to the States in April, shortly after Roosevelt's death, but Truman bluntly refused.* Quesada, instead, became the chief intelligence officer of the Army Air Forces, "an administrative job that just bored the hell out of me." After Japan capitulated later in the year, Quesada, still in the new Pentagon, became part of an informal group including Spaatz, Eaker, Fred Anderson, Lauris Norstad, and Hoyt Vandenberg, which set out to sell the idea of a separate air force. While negotiations with the Navy took place at the Secretary's level, this group worked to persuade senators and the Army of the soundness of the plan. Quesada's role was to convince the Army, specifically Eisenhower and Bradley, and Senator Leverett Saltonstall, the Chairman of the Armed Services Committee, that the Army did not need its own tactical air force. Principally through his wartime relationship with them, Quesada persuaded the Army generals that the air force knew better how to use its airplanes and that the flexibility air power had demonstrated so successfully in the war would be maintained by a separate air force. At one point, with Quesada present, Spaatz promised Eisenhower that if Eisenhower supported separation, the Air Force would always meet its commitment to the Army by providing permanent and strong tactical air forces. In part as a result of this promise, Eisenhower and Bradley were won over.

When it appeared that separation was assured the following March, Quesada took over the Third Air Force in Tampa, Florida. At the same time the postwar Army Air Forces was divided into three separate but unequal branches: Strategic Air Command, Air Defense Command, and Tactical Air Command. Two months later, Quesada started building the Tactical Air Command combining his Third Air Force, the old Ninth and Twelfth Air Forces, and the wartime IX Troop Carrier Command. The wartime arrangement whereby numbered air forces were made up of commands was reversed, and commands were now composed of numbered air forces. The new Tactical Air Command's mission was to be prepared to participate in joint operations with the Army and Navy to perform interdiction operations on its own.

Quesada approached his new job with the conviction that the best way to keep the tactical air mission from falling back under the Army was to provide such outstanding support that the Army would be totally satisfied and forget about having its own air force. In this he was fully supported by his Plans and Operations officer, Col. William Momyer. In May, Quesada transferred his headquarters to Langley Field and in October was joined next door at Fort Monroe by Gen. Jacob L. Devers and his Army Ground Forces Headquarters. At that time Quesada received his third star. The two commanders set about to institutionalize their wartime experiences in air-ground cooperation.

Quesada's attempt to build a combat arm was hampered at first by the need to offset the rapid postwar demobilization with a vigorous recruiting program. On top if this, Truman embarked on a severe budget-cutting program that forced the AAF to lower its plans for expansion from seventy to fifty-five

---

* He had a military aide—an Army ground officer.

246 · AVIATION QUARTERLY · Spring 1989

TAC AIR COMES OF AGE

Top-ranking officers touring American installations in Europe, l. to r.: Lt. Gen. Omar Bradley, Gen. Hap Arnold, Gen. Dwight Eisenhower, and Gen. George Marshall. *(Photo: The Air Force Historical Foundation)*

groups of airplanes. By year's end TAC manpower had dwindled drastically. Its three numbered Air Forces were cut to two with inactivation of the Third Air Force. The number of troop carrier wings that had been planned was reduced and several airfields closed. In a country weary of war, the AAF was experiencing difficulty attracting good people into the aviation cadet program.

Undaunted, Quesada pressed on with establishing the command. In November 1946, he introduced the first jet plane, the P-80, as a successful fighter-bomber for close air support. By August 1948, he had integrated the first F-84s into Tactical Air Command. Throughout 1947 and 1948, he and Devers experimented with joint training. The Army wanted to split the planes up into small groups to work directly with individual ground units. Quesada, ever mindful of the flexibility that had served him so well during the war, convinced first Devers and then Spaatz that the tactical airplanes should remain centrally managed by him. Late in 1947, the Twelfth Air Force practiced supporting amphibious landings off the coast of Florida and trained with the 2d Infantry Division in Alaska. At the same time, the Ninth Air Force dropped airborne soldiers of the 82d Airborne Division in New York State. The following March, the Twelfth Air Force worked with Army ski troops in the mountains of Colorado, and during May provided column cover to the tanks of the 2d Armored Division in Texas. Quesada's wartime experiences were finally being translated into doctrine and practice.

The starkest illustration of Quesada's theories of air-ground cooperation was a week-long exercise, called Operation Combine, in which all fighter, reconnaissance,

MAKERS OF THE UNITED STATES AIR FORCE

and troop carrier squadrons, several units of the Strategic Air Command, and elements of the 82d Airborne Division demonstrated the awesome power of combined arms. The Ninth Air Force put on this demonstration each year at each of the eight Army Ground Forces schools. Along with this training in close air support, the Ninth Air Force practiced interdiciton by repelling a hypothetical invasion off the Carolina coast. By means of these exercises Quesada was honing not only the airmen but also tactical air doctrine that had flowed from World War II. By 1948 the command had 300 fighters, half of them jets; 63 medium bombers; 100 reconnaissance planes, also half jets; 237 airlift planes; and 80 liaison aircraft. During the summer Quesada sent several thousand traffic controllers and maintenance specialists to Europe to help break the Soviet ground blockade of Berlin.

While Quesada battled with elements in both the newly created Air Force, and the Army, which wanted to create its own tactical air arm, worldwide developments were conspiring against his efforts. Since he took over the command in May 1947, the Cold War had set in. That very month Hungary had installed a Communist government, followed in June by the announcement of the Marshall Plan. Early in 1948, Czechoslovakia followed Hungary, and in the summer the Soviets tried to cut off Berlin from the Allies. In June, Hoyt Vandenberg replaced Spaatz as Air Force Chief of Staff. The emphasis on strategic preparedness and deterrence, which had been instrumental in creating the separate Air Force, assumed even greater significance. In the fall of 1948, Quesada was called to Washington and informed by Vandenberg that he was going to reorganize the Air Force's operational commands. The Strategic Air Command would be strengthened while the Tactical Air and Air Defense Commands would be reduced to headquarters and placed under a new Continental Air Command. Quesada objected, reminding the chief of the promise to Eisenhower that there would always be a tactical force to support the Army. Vandenberg, disagreeing with Quesada's philosophy that

Northrop's P-61 Black Widow night fighter saw combat late during WWII.
(Northrop Photo)

TAC AIR COMES OF AGE

the best way to keep tactical air out of Army hands was to make it indispensable while under the Air Force, and viewing Quesada's attempts at cooperation as a pathway to Army domination, went ahead with the plan. Spaatz, now retired, was furious. Quesada, "personally offended" at what he considered a violation of trust, turned down an offer to head the new Continental Air Command. Instead, he went to the Pentagon to help draft legislation aimed at nationalizing the Air National Guard, an assignment he later characterized as "very unpleasant, disagreeable, and unsuccessful."

In 1949, Quesada's diplomatic skill was again called upon as he headed the nation's first hydrogen bomb test at Eniwetok Atoll in the Pacific. Working for both the Atomic Energy Commission and the Defense Department, he first weeded out from amidst hundreds of suplicants the experiments to be included. He then commanded the test, which involved building the site, installing the equipment, negotiating with scientists from Robert Oppenheimer to Edward Teller, and seeing to the smooth running of the project. The test was a success in large part due to Quesada's skill in organizing and mollifying.

Amidst a swirl of media reports that he was "resigning in protest" over the treatment of tactical air power, Quesada retired in 1951. If disillusionment and disappointment over the broken promise to Eisenhower did not cause this decision, they at least accompanied it. Having been married four years earlier to Kate Davis Pulitzer, daughter of St. Louis publisher Joseph Pulitzer, and by then the father of a young son, he felt impelled to embark on a civilian career. His technical and managerial abilities resulted in successful stints as a manager at Olin Industries, organizer and director of Lockheed's Missile System Division, special assistant to President Eisenhower, administrator of the Federal Aviation Administration, part owner of the Washington Senators baseball team, and president of L'Enfant Properties in his native Washington, D.C.

Quesada continued his crusade for stronger interservice cooperation after his retirement. In an article in *Colliers* magazine in 1956, he became the first in a line of critics, including one later Chairman of the Joint Chiefs of Staff, to recommend publicly stronger unification of the services and reorganization of the Joint Chiefs. In particular he proposed, as have others since, that the Joint Chiefs no longer remain heads of their individual services when entering the joint arena.

USAF's first operational jet was the Lockheed P-80 shooting star. *(USAF Photo)*

The Tactical Air Command was revived during the Korean conflict and strengthened during the 1960s as the nation shifted to a military strategy of flexible response and fought the war in Vietnam. Both the doctrine and tactics of America's reborn strong tactical air arm mirrored Quesada's accomplishments in World War II and in his postwar creation of the command. The Air Force acknowledges the debt it owes to this pioneer. Its pantheon of "senior statesmen" who meet each year with the Air Force Chief of Staff is composed basically of retired four-star generals—Ira Eaker, Jimmy Doolittle, and Elwood "Pete" Quesada. Series Cont'd: AQ Volume 9 - Number 3.

MAKERS OF THE UNITED STATES AIR FORCE

Above: F-86 by North American Aviation. The first jet to go supersonic in level flight. *(USAF Photo)*
Below: Boeing B-52 with chute deployed. *(USAF Photo).*

Author

Prior to his retirement from the Air Force in September 1983, COLONEL JOHN SCHLIGHT was Deputy Chief of the Office of Air Force History. A master navigator, Colonel Schlight flew in the Korean War and in Indochina during the French conflict with the Viet Minh in the early 1950s. Later assignments included a tour in Vietnam, duty as a member of the Air Force Academy Department of History, and Dean of Faculty and Academic Programs at the National War College. He holds a doctorate in medieval military history from Princeton University and is the author of two books on military history of the Middle Ages. He also has written extensively on Air Force operations in Vietnam. Colonel Schlight is presently Chief of the Southeast Asia Branch at the U.S. Army's Center of Military History.

Editor's note:
More about John Schlight on page 256

Sources

General Quesada's papers reside principally at two locations: the Manuscripts division of the Library of Congress in Washington D.C., and the Dwight D. Eisenhower Library in Abilene, Kansas. The former collection contains correspondence and official records relating to World War II. The majority of the papers at the Eisenhower Library deal with Quesada's post-military career in civilian aviation. A relatively small number of military papers there consist of the general's wartime correspondence with the families of airmen and of documents concerning his tenure as Commander of the Tactical Air Command. A limited number of speeches and interviews can be found at the USAF Historical Research Center at Maxwell AFB, Alabama.

Four oral history interviews with the general are extant. One done in 1960 for the *American Heritage* journal and another with the Office of Air Force History in 1975 include many details of both his military and civilian careers. A third interview, with the author in 1982, confines itself to his military years. Finally, the Office of Air Force History has published a group interview, *Air Superiority in World War II and Korea* (Office of Air Force History/GPO, 1983), in which Quesada discusses the question of air superiority with three former colleagues in the Tactical Air Command, Generals William W. Momyer, Robert M. Lee, and James Ferguson.

Quesada's views on air-ground and interservice cooperation are summarized in several presentations he made to a group of Air Force and Army officers in 1947 and 1948 and in a *Colliers* magazine article published in 1956 titled "Peace at the Pentagon."

Several official histories of the Ninth Air Force's operations in Europe in 1944 exist at Maxwell AFB and in the Office of Air Force History in Washington, D.C. These repositories also contain histories of the Tactical Air Command and of many of the fighter groups which Quesada led in World War II.

General Quesada makes cameo appearances in many of the memoirs by American military leaders of World War II, including Eisenhower, Bradley, and Arnold. He is also treated briefly in such secondary works as Kenn C. Rust's *The 9th Air Force in World War II* (Aero Publishers, 1967); Kent Roberts Greenfield's *American Strategy in World War II: A Reconsideration* (John Hopkins Press, 1963); Russell F. Weigley's *Eisenhower's Lieutenants* (University of Indiana Press, 1981); and DeWitt S. Copp's *A Few Great Captains* and *Forged in Fire* (Doubleday, 1980 and 1982). AQ

(MOTUSAF continues in AQ Volume 9, Number 3, Summer 1989)

Convair YB-60, which lost out to Boeing's B-52. Only one YB-60 built in 1950. *(Photo: AQ files)*

MAKERS OF THE UNITED STATES AIR FORCE

A Stinson L-5 makes a drop to G.I.s at a forward gun position during "The Forgotten War," Korea. *(USAF)*

The little known McDonnell XF-85 parasite fighter. Carried by the B-36, but it wasn't feasible. *(USAF Photo)*.

AVIATION QUARTERLY

German ME262 two-seater trainer, which was also experimenting with night intercepts near end of WWII. This Aircraft was captured by English troops in early May 1945.
(Photo: Smithsonian Institute)

AVIATION QUARTERLY

P-47 Thunderbolt is the pride of Republic. Designed and modified to accompany high-altitude daylight bombers on long-distance raids, its eight machine guns make it a fine strafer and its power makes it a good fighter bomber.

AVIATION QUARTERLY

German ME262 two-seater trainer, which was also experimenting with night intercepts near end of WWII.
This Aircraft was captured by English troops in early May 1945.
(Photo: Smithsonian Institute)

AVIATION QUARTERLY

P-47 Thunderbolt is the pride of Republic. Designed and modified to accompany high-altitude daylight bombers on long-distance raids, its eight machine guns make it a fine strafer and its power makes it a good fighter bomber.

AVIATION QUARTERLY

P-38 Lockheed Lightning has been revised to increase range and altitude. New aileron boosters improve the plane's climb and maneuverability. Besides its cannon and four machine guns, it now can carry two 1,000-lb. bombs.

AVIATION QUARTERLY

The United States Air Force in Southeast Asia
THE WAR IN South Vietnam
The Years OF THE Offensive 1965·1968

John Schlight

To order:
Call 202-783-3238 or write: Supt. of Documents, Government Printing Office, Washington, D.C. 30402-9325. Credit cards or checks acceptable.

Involved with the conflicts in Southeast Asia throughout his 31-year Air Force career, Col. John Schlight, USAF, Retired, flew aircraft in Indochina in support of the French during the early 1950s, was deputy director of the Air Force's Project CHECO (Contemporary Historical Evaluation of Combat Operations) in South Vietnam in 1969 and 1970, and headed the Vietnam War Section in the Office of Air Force History from 1977 to 1981. With a PhD in history from Princeton University, he has taught military history at the United States Air Force Academy, the National War College, and at universities in the United States and overseas. He is the author of *Monarchs and Mercenaries* and *Henry II Plantagenet* and editor of *The Second Indochina War*. Colonel Schlight's last assignment with the Air Force was Deputy Chief, Office of Air Force History. He is currently Chief of the Southeast Asia Branch at the U. S. Army's Center of Military History in Washington, D.C.